Technical Rescue Field Operations Guide

How to use this guide

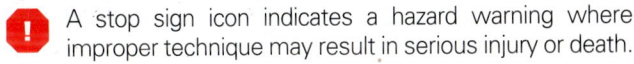 A stop sign icon indicates a hazard warning where improper technique may result in serious injury or death.

✓ Text with a red check next to it denotes a key part of that specific procedure.

✓ Text with a red check and colored red denotes a critical aspect of that procedure that must not be overlooked.

Red rope in this guide denotes working line.

Belay line has no fill or is grey.

On checklists:

☐ A black box indicates an operational level skill
☐ A red box indicates a technician level skill

✓ Remember that this is only a guide. The rescuer must adapt to each situation as common sense dictates!

✓ Always be sure the system passes the whistle test, which means if all rescuers were to let go of the systems, no catastrophic failure or injuries would occur.

✓ Always be sure all personnel tie in near the edge.

*Metric measurements are included in parentheses in the rope rescue section and are converted to the nearest centimeter or half meter.

All product names throughout this handbook are trademarks of their respective holders.

NFPA 1670 Standards

NFPA 1670 1999 edition is the current standard for technical rescue operations. Skills and procedures in this guide are categorized according to NFPA 1670 standards where applicable. For example, a command checklist involves scene size-up and hazard recognition. This is an awareness level skill and will have an **Ⓐ** icon. High angle litter raising is a technician level skill and will have a **Ⓣ** icon.

Some procedures are not specifically addressed in the NFPA standard and the categorization is interpreted from similar categorized procedures. An interpreted level will have an asterisk next to the icon.

The intent is to make it easy for responders to assess different situations appropriately and to train according to standardized guidelines.

The general definitions of the 1670 operational levels are in Appendix A. Refer to the NFPA 1670 document for complete definitions in each area and for each skill or procedure.

The authority having jurisdiction has final say as to the categorization of each procedure.

Ⓐ Awareness level skill   **Ⓞ Operational level skill**

Ⓣ Technician level skill

NFPA 1983, 2001 Edition

NFPA 1983 is the U. S. Fire Service standard for life safety rope and harnesses. It defines all rescue system components, their construction, use, labeling and testing.

A two person system icon means that the procedure is intended for two person loads and any component that will bear the weight of two people must be rated for general use. Light duty components may be part of the system when they are used to support the weight of a single person.

Ⅱ Two person (general use) components

Contents

Key procedures in red

Contents

Key procedures in red

Risk Management

Safety is always our first concern

At the start of each operation, ask these questions;

1. What is the key problem?
2. What is our plan of action?
3. Why is that the safest plan?
4. What are the biggest risks that we need to watch out for?
5. What is your gut feeling about this plan?

Remember

- We will risk our lives in a calculated way that is appropriate to the situation to save savable lives
- We will not risk our lives at all for that which is already lost

Communicate

- Each operation must have a clearly defined leader
- A decision on rescue or recovery strategy must be made clear to everyone at the outset of every operation
- Speak up if you see a problem no matter how small or obvious it may seem

Re-evaluate strategy whenever appropriate

- When new information becomes known
- When a significant event occurs
- After an extended time period has elapsed

Incident Management

Most technical rescue incidents are focused around a small number of subjects and can be easily handled with a simple Incident Command System (ICS) structure. Overall command can be any officer but a rescue technician should assume an operations level role and manage the technical rescue portion of the incident.

Each discipline has a specific command checklist with key tactical benchmarks. Use the checklist.

First Responders
- Take command and size up
- Focus on information gathering
- Identify hazards
- Be certain that the right resources are called early
- Avoid activities marked in red on checklists and in text

Rescue Technician/Operations Officer (TSO)
- Assume operations control
- Review hazards and critical factors
- Assist with the formation of incident action plan and backup plan
- Assign sectors and deploy resources
- Keep command informed about all phases of the operation
- Communicate with sectors and revise plans as needed

Rescue Technician/Sector Officer (i.e. Rescue Sector)
- Clearly understand the action plan
- Communicate the action plan to sector personnel
- Supervise task level activities
- Keep Operations Officer or Incident Commander updated on a regular basis

Technical Sector Officer (TSO) is normally a term for the person in charge of a group or sector (i.e. Rescue Sector). In some cases the TSO may function as the overall rescue leader.

Incident Management

Mountain Rescue example

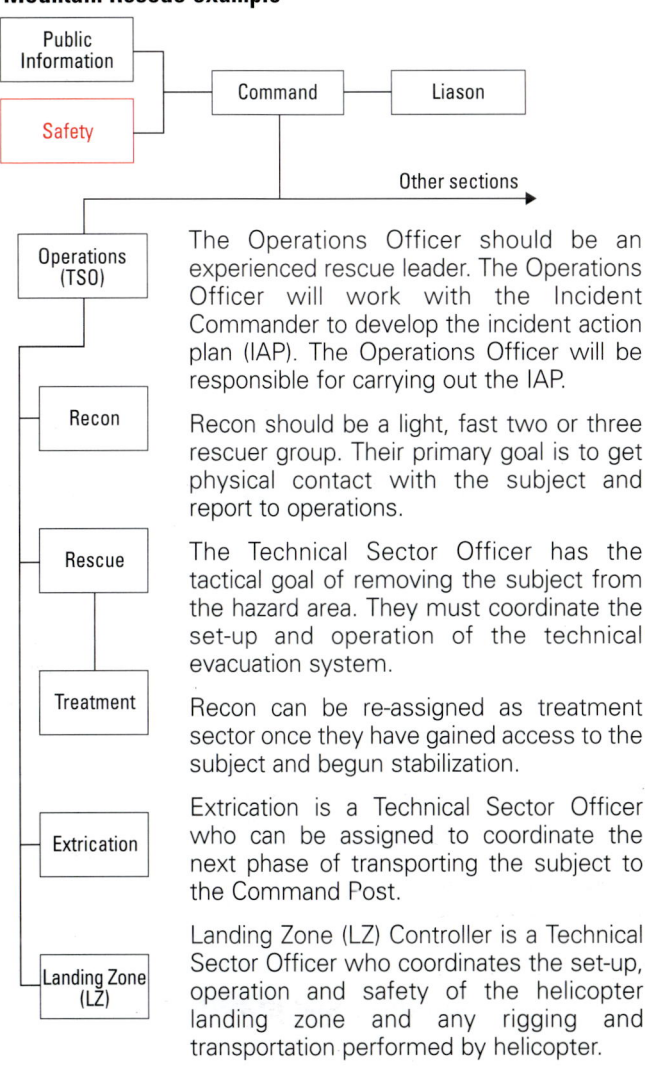

```
Public
Information

Safety

        Command ──── Liason

                Other sections

Operations
(TSO)

Recon

Rescue

Treatment

Extrication

Landing Zone
(LZ)
```

The Operations Officer should be an experienced rescue leader. The Operations Officer will work with the Incident Commander to develop the incident action plan (IAP). The Operations Officer will be responsible for carrying out the IAP.

Recon should be a light, fast two or three rescuer group. Their primary goal is to get physical contact with the subject and report to operations.

The Technical Sector Officer has the tactical goal of removing the subject from the hazard area. They must coordinate the set-up and operation of the technical evacuation system.

Recon can be re-assigned as treatment sector once they have gained access to the subject and begun stabilization.

Extrication is a Technical Sector Officer who can be assigned to coordinate the next phase of transporting the subject to the Command Post.

Landing Zone (LZ) Controller is a Technical Sector Officer who coordinates the set-up, operation and safety of the helicopter landing zone and any rigging and transportation performed by helicopter.

Time Management

Time is a critical factor. History has proven that performing tasks sequentially to accomplish the objective consumes the greatest amount of time.

Tips for a safe and fast rescue

- The Technical Sector Officer (TSO) has the big picture, coordinating and fine tuning all parts of the technical evacuation sector
- Multitasking with simultaneous performance of tasks is the goal
- Individuals must work as quickly as possible to accomplish their task, but must not compromise safety for speed
- The TSO must avoid performing hands-on tasks in order to retain overall control of the sector
- Frequent operation specific training is necessary for a safe, effective and efficient team

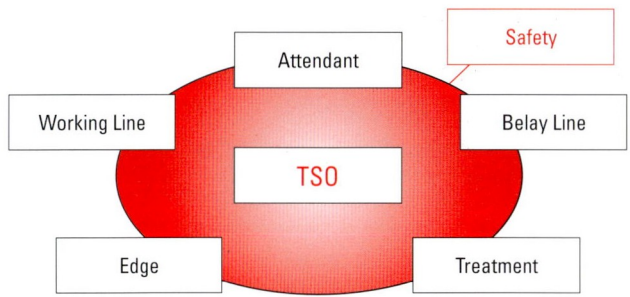

Rope Rescue Command Checklist

Phase I: Size up

☐ **Primary assessment**
- ☐ Secure witness or reporting party (RP)
- ☐ Determine location, number and condition of victims
- ☐ Identify hazards to rescuers (rock fall, terrain etc.)
- ☐ Choose rescue mode or recovery mode

☐ **Secondary assessment**
- ☐ Type of terrain
  - ☐ Non-technical (<40°)
  - ☐ Technical (>40°)
- ☐ Assess the need for additional personnel and or equipment (helicopter, support truck)

Phase II: Pre-rescue operations

☐ Make general area safe (i.e., traffic and crowd control)
☐ Make rescue area safe
- ☐ Establish lobby control and accountability
- ☐ Designate safety officer
- ☐ Develop incident action plan (see decision tree p. 10)
- ☐ Develop backup plan

☐ Proper personal protective equipment
☐ Appropriate rescue and patient packaging equipment
☐ Equipment for subject (helmet, water, eye protection)
☐ Pre-rescue briefing

Phase III: Rescue operations

☐ Deploy personnel
- ☐ Insertion technique: hike, climb, helicopter, longline
- ☐ Evacuation technique
  - ☐ Low angle, high angle raise/lower
  - ☐ Helicopter, internal load or longline

☐ Transfer to Advanced Life Support (ALS)

Phase IV: Termination

☐ Removal of equipment
☐ Personnel Accountability Report (PAR)

A

Personal Protective Equipment

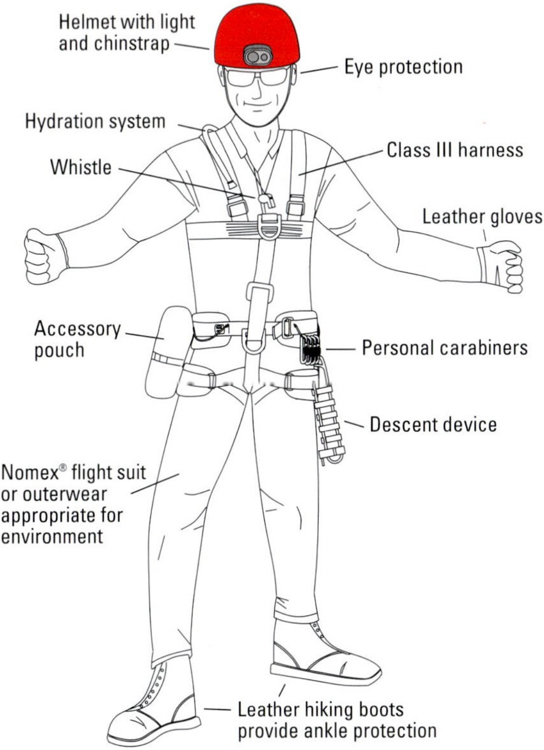

Helmet with light
and chinstrap

Eye protection

Hydration system

Whistle

Class III harness

Leather gloves

Accessory
pouch

Personal carabiners

Descent device

Nomex® flight suit
or outerwear
appropriate for
environment

Leather hiking boots
provide ankle protection

Accessory pouch
- ☐ Personal purcells
- ☐ 15 ft. (4.5m) webbing
- ☐ Extra batteries
- ☐ Energy food
- ☐ Trauma shears

Radio harness
(not pictured)
- ☐ Portable radio
- ☐ Pen
- ☐ Paper

Terrain Types

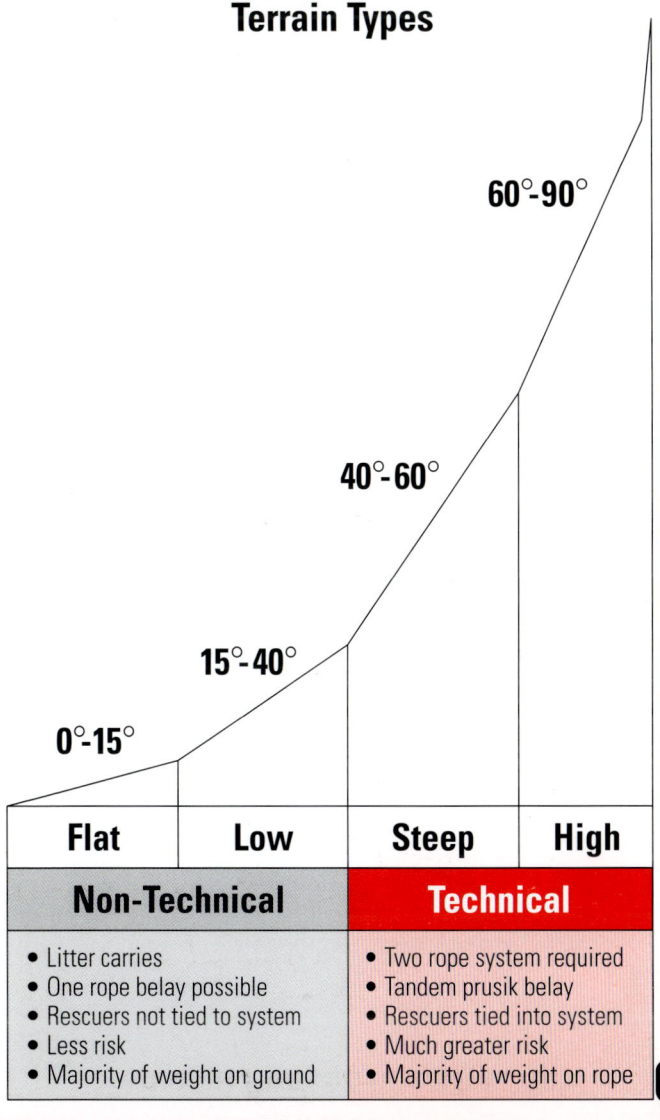

60°-90°

40°-60°

15°-40°

0°-15°

Flat	Low	Steep	High
Non-Technical		**Technical**	
• Litter carries • One rope belay possible • Rescuers not tied to system • Less risk • Majority of weight on ground		• Two rope system required • Tandem prusik belay • Rescuers tied into system • Much greater risk • Majority of weight on rope	

A

Mountain Rescue Decision Tree

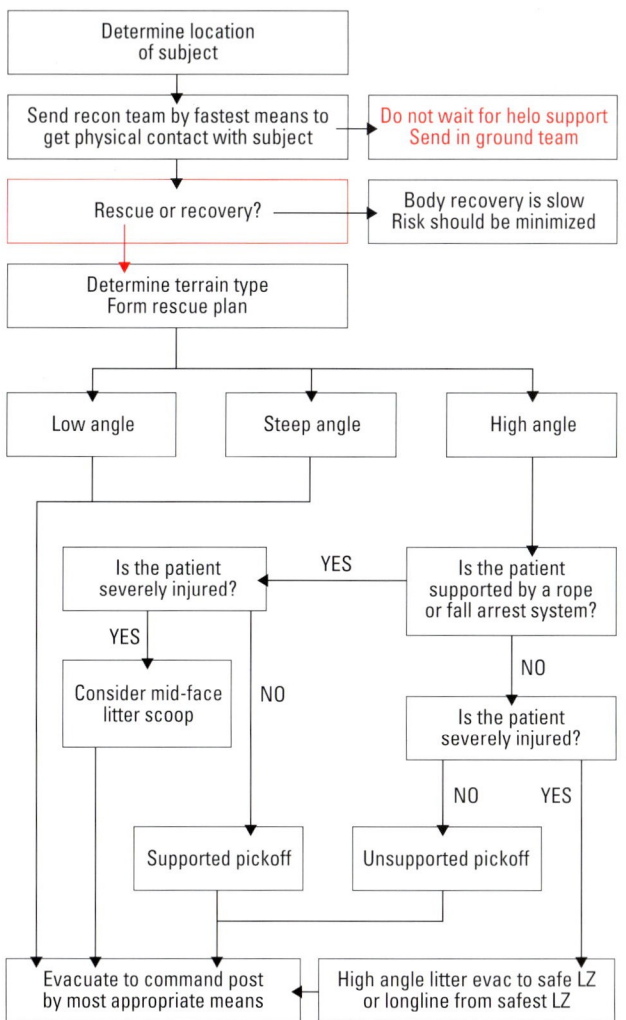

Basic Life Safety Knots

Figure Eight on a Bight (end of rope anchor knot)

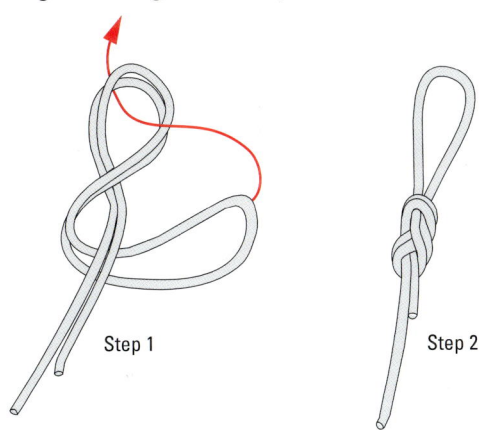

Step 1

Step 2

Figure Eight Follow Through (tie off for harness or anchor point)

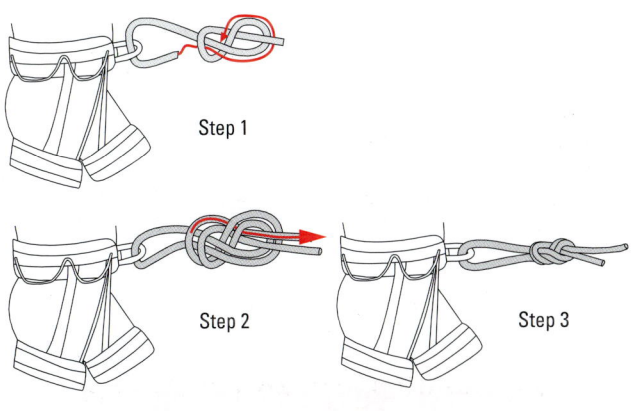

Step 1

Step 2

Step 3

Basic Life Safety Knots

Bowline (end of rope anchor knot)

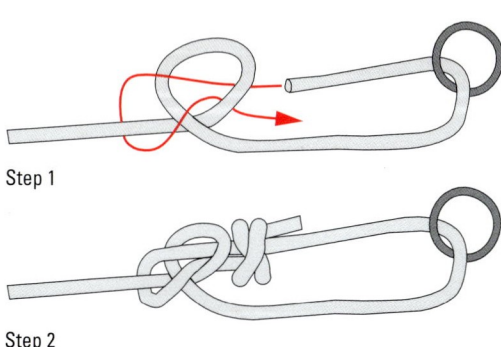

Step 1

Step 2

Butterfly (middle of rope knot)

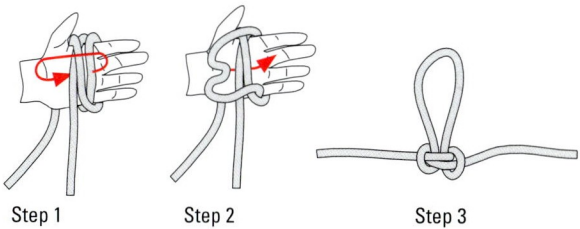

Step 1   Step 2   Step 3

Basic Life Safety Knots

Double Overhand Bend (tie two ropes together)

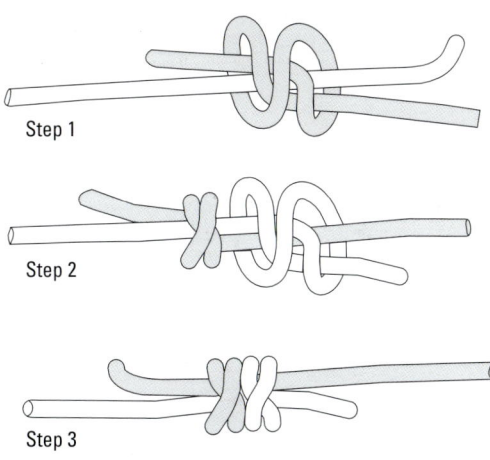

Step 1

Step 2

Step 3

Water Bend (tie webbing together)

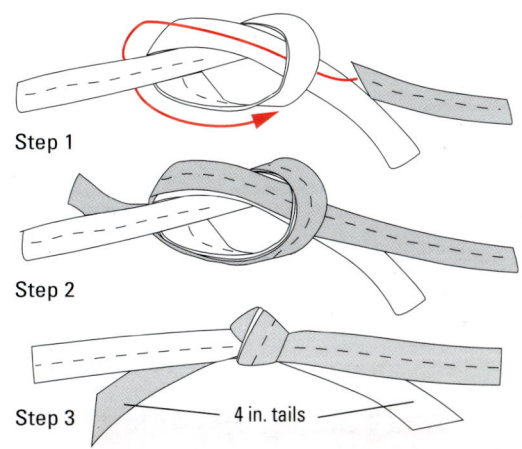

Step 1

Step 2

Step 3   4 in. tails

Basic Life Safety Knots

Prusik Hitch

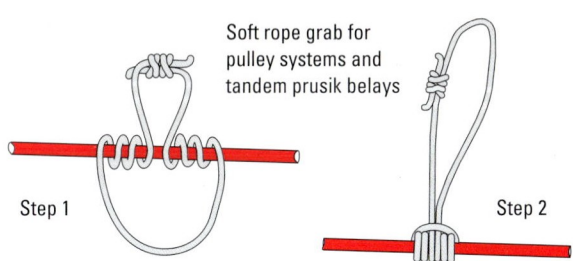

Soft rope grab for
pulley systems and
tandem prusik belays

Step 1   Step 2

Used in pairs, 54 in. (137 cm) and 66 in. (168 cm), prior to tying
for tandem prusic belays

Münter Hitch

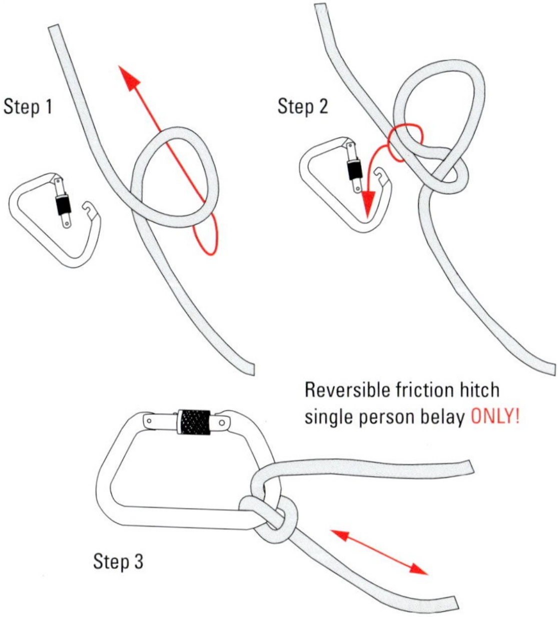

Step 1

Step 2

Reversible friction hitch
single person belay ONLY!

Step 3

Load Releasing Hitch (LRH)

- Component of tandem prusik belay
- Used for knot passing
- Made with 33 ft. (10m) 9mm rope and two steel carabiners

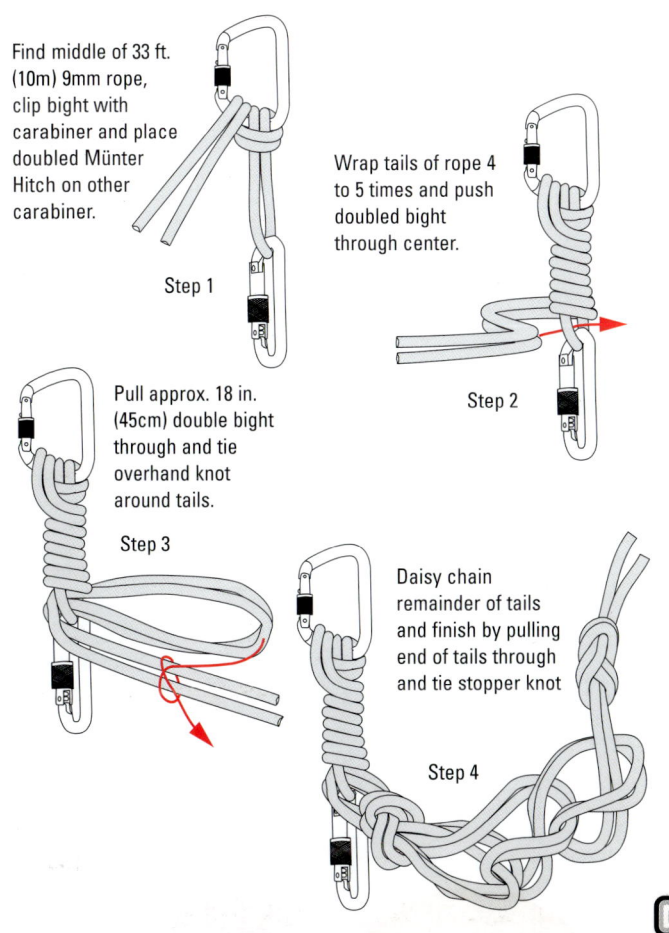

Find middle of 33 ft. (10m) 9mm rope, clip bight with carabiner and place doubled Münter Hitch on other carabiner.

Step 1

Wrap tails of rope 4 to 5 times and push doubled bight through center.

Step 2

Pull approx. 18 in. (45cm) double bight through and tie overhand knot around tails.

Step 3

Daisy chain remainder of tails and finish by pulling end of tails through and tie stopper knot

Step 4

Personal Purcell Prusik System

Uses
- Ascending a fixed line
- Self rescue
- Team based pick-offs
- Litter attendant tie in
- Travel restraint near edge for edgemen and spotters

Construction
- 33 ft. (10m) 6mm accessory cord
- Size anatomically for individual as shown on sizing illustration
- Dimensions shown are for tied and finished purcells not hitched to rope

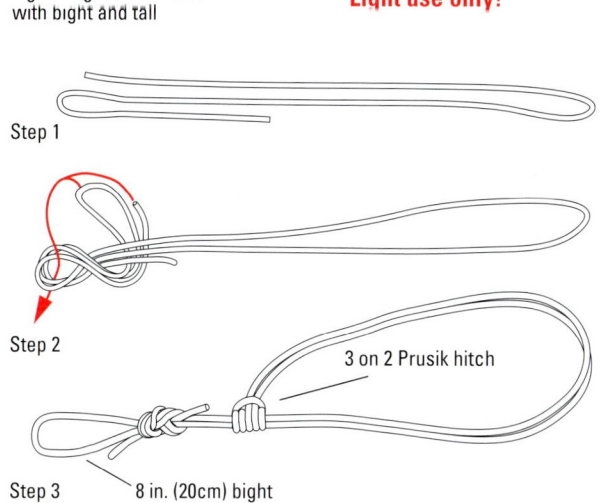

Figure eight bend made with bight and tail

Light use only!

Step 1

Step 2

3 on 2 Prusik hitch

8 in. (20cm) bight

Step 3

Sizing Personal Purcells

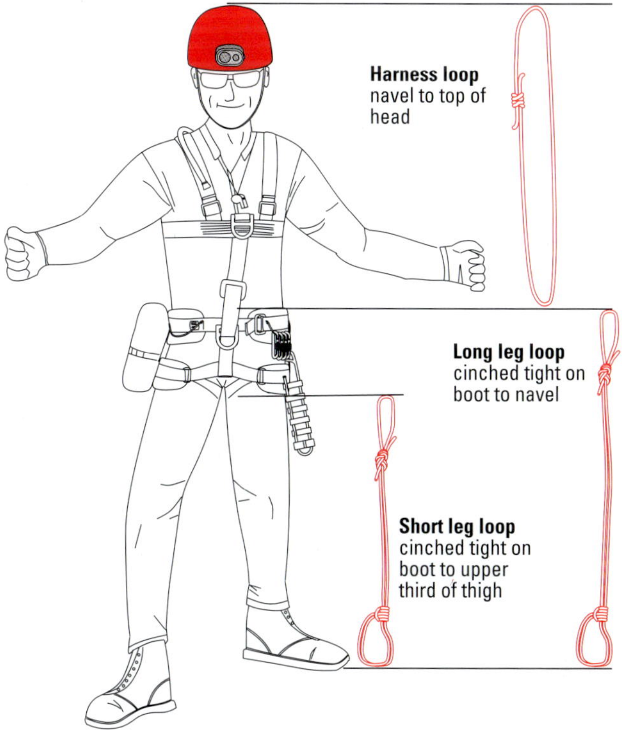

Harness loop
navel to top of head

Long leg loop
cinched tight on boot to navel

Short leg loop
cinched tight on boot to upper third of thigh

✓ These prusik lengths are approximate and will vary from individual to individual based on anatomical reference points and personal preference. The harness loop must be long enough to prusik onto the rappel line above the descent device and clip into your harness.

Self Rescue

A rescuer must always be prepared to perform self rescue procedures, specifically the ability to ascend a fixed rope, free a jammed descent device, pass a knot on rappel or any combination of the above.

! Warning! The belay rope has been left out of this illustration for clarity and also to make the point that in a self rescue situation a belay might not be available. *Extreme* caution is warranted when there is no belay. An option is to tie a bight into the rope below the leg purcell and clip it to the harness. Tie a new bight every 5 linear ft. (1.5m) of rope climbed and clip to harness.

✓ Always use a second rope belay whenever possible!

Ascend fixed rope

1. Attach harness prusik loop to rope with prusik hitch.
2. Connect harness loop to harness with locking carabiner.
3. Attach climbing purcell to rope with prusik hitch.
4. Place foot into small loop and cinch tight onto foot.
5. Sit back onto harness prusik and move climbing prusik up toward harness prusik with free hand.
6. Stand up onto climbing prusik and move harness prusik up as far as possible.
7. Repeat this process until reaching destination or until problem is solved.

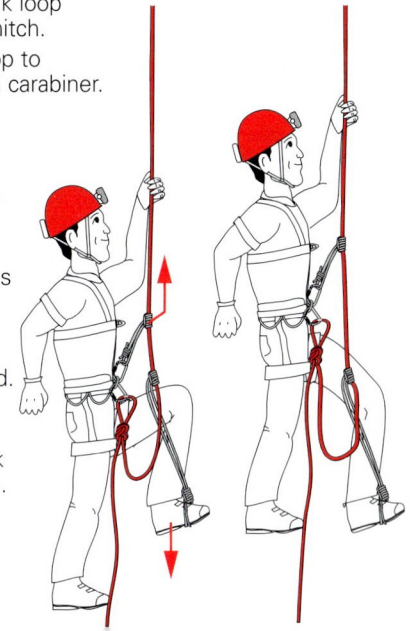

Patient Packaging

The following five points should be patient packaging goals.

1. Immobilize the patient to minimize movement no matter what position the litter is placed in.
2. Use plenty of padding under the patient and in all voids.
3. Protect the head and face from debris and vegetation.
4. Protect the patient from the elements both hot and cold.
5. Give special consideration to vital signs and airway management.

Leave buffer space between the patient's head and the end of the litter

15 ft. (4.5m) Red webbing has no tension. Used as restraint precaution to prevent patient's head from pushing against end of litter.

20 ft. (6m) webbing for seat and upper body

30 ft. (9m) webbing for feet and lower body

Pad all void areas

Modify packaging if needed for injured extremities

Commercially made strap systems also work well

✓ This illustration is but one example of a number of techniques that can be used.

Cold weather

Use a three layer system.
1. Vapor barrier against skin.
2. Insulation layer.
3. Weather barrier.

Heat packs must be added if the patient is hypothermic or if the evacuation will take longer than 30 minutes.

Low Angle Evacuation 0° - 40°

- Majority of weight on ground
- 4 to 6 bearers
- Litter wheel optional depending on terrain
- 15 ft. (4.5m) carry strap for each bearer optional
- Single belay line optional depending on terrain
- Use tandem prusik for belay
- Do not load belay line or use to lower

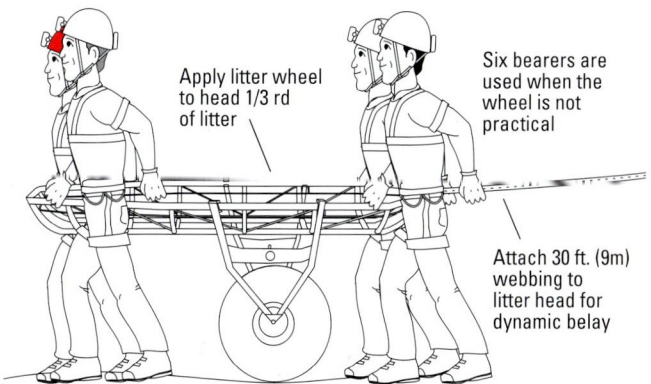

Apply litter wheel to head 1/3 rd of litter

Six bearers are used when the wheel is not practical

Attach 30 ft. (9m) webbing to litter head for dynamic belay

Caterpillar pass

- Consider caterpillar pass to negotiate short sections of technical terrain
- Get solid footing and stay in position while passing litter hand to hand
- Very effective on short sections of steep and high angle obstacles
- This is an effective technique but at least 10 or 12 bearers are desirable to accommodate personnel rotation
- Consider belay line

Anchor Systems

Definitions

- Natural anchors: naturally occurring trees and rocks
- Artificial anchors: anything placed by man including fire trucks and structural members
- Bombproof anchor: an anchor that you confidently believe will hold the intended load and any potential impact force unintentionally generated by the load
- Marginal anchor: an anchor that you do not believe to be bombproof
- Single-point anchor: single point of origin
- Multi-point anchor: a collection of marginal point anchors connected into a bombproof anchor system
- Back-tie anchor: a marginal anchor in a good location that is linearly connected with a tensioning unit to a bombproof anchor somewhere back from the edge

Concepts

- Safety test all anchors in the direction of use with a force comparable to the working load
- Watch for signs of weakness or failure
- Distribute force equally between all anchors in a multi-point system
- On multi-point anchors, keep the distributing link small to minimize any potential impact force
- Always try to have independent anchors for the working line and belay line
- Choose strong points like joints and corners on structural members for anchors
- Avoid mid-span anchor points on structural members if possible
- When using pre-sewn straps, prevent side loaded or tri-loaded carabiners

Anchor Systems

- Keep angle less than 90°
- Commercially made straps are acceptable
- Commit entire ropes to the anchor if necessary
- Pad all sharp edges

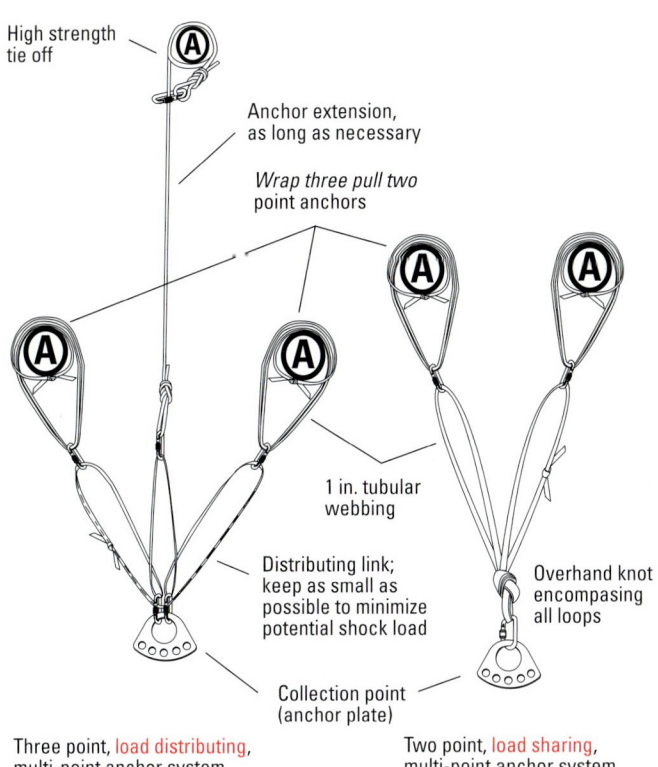

High strength tie off

Anchor extension, as long as necessary

Wrap three pull two point anchors

1 in. tubular webbing

Distributing link; keep as small as possible to minimize potential shock load

Overhand knot encompassing all loops

Collection point (anchor plate)

Three point, load distributing, multi-point anchor system (potential shock-load)

Two point, load sharing, multi-point anchor system (minimizes shock-load)

✓ Always double-check everything!

Back-Tie Anchors

A back-tie anchor is used to focus a marginal anchor to a bombproof anchor. It is built with low-stretch rope and a system prusik.

1. Construct back-tie system as shown with ratchet prusik on line closest to haulers.
2. Three wraps of 1/2 in. (13mm) rope is ideal but distance between anchors and available rope may limit number of wraps.
3. Pull tension on system and set ratchet so that all ropes stay under tension but do not damage forward anchor.
4. Tie off back-tie tension unit.

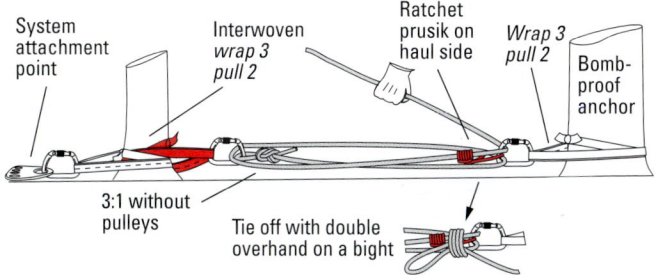

System attachment point

Interwoven *wrap 3 pull 2*

Ratchet prusik on haul side

Wrap 3 pull 2

Bomb-proof anchor

3:1 without pulleys

Tie off with double overhand on a bight

✓ Keep back-tie anchor in line with the fall line.

✓ As a rule, look for bombproof single-point anchors and linear anchors for rescue.

✓ Multi-point anchor systems made of marginal anchors should be the rare exception.

Directional Anchors

The fall line refers to the natural plumb line always present as a result of gravity. It is affected and changed by the angle and aspect of the slope.

The location of suitable anchors relative to a suitable fall line is always a critical factor. Occasionally, it is safest to redirect the system into a directed fall line with a directional anchor.

Caution: a change of direction can place up to 200% of the load on the directional anchor depending on the vector angle. A 90° vector angle will place approximately 140% of the load on the directional anchor.

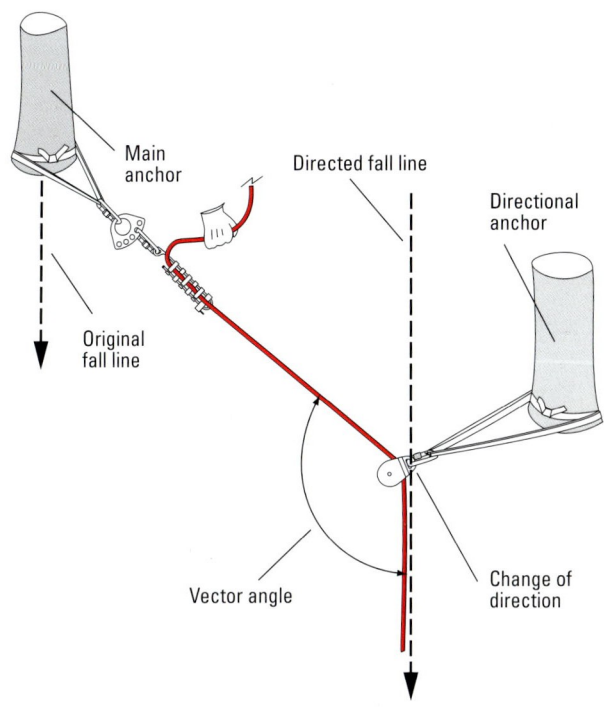

Main anchor

Directed fall line

Directional anchor

Original fall line

Vector angle

Change of direction

Structural Anchors

Pre-sewn anchor straps
- Fast to set up
- No knots
- Rated auxiliary equipment

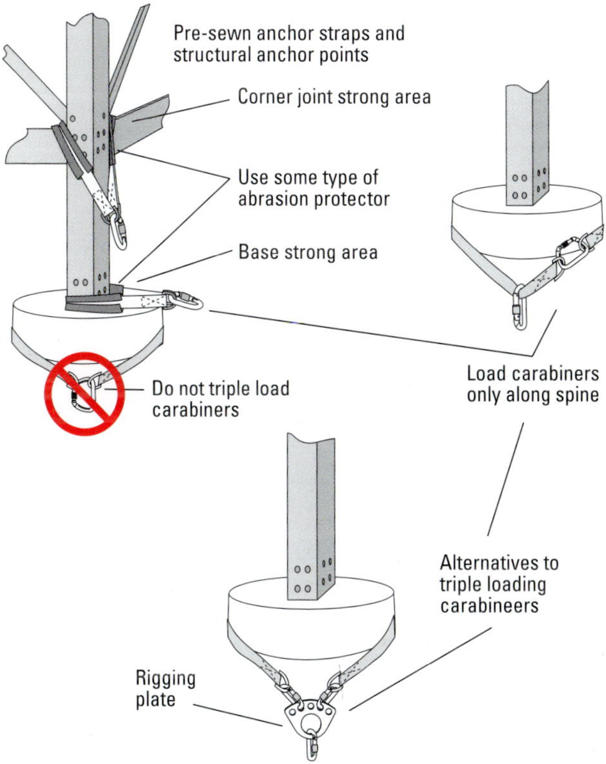

Pre-sewn anchor straps and structural anchor points

Corner joint strong area

Use some type of abrasion protector

Base strong area

Do not triple load carabiners

Load carabiners only along spine

Alternatives to triple loading carabineers

Rigging plate

Fixed Belay for Edgemen

Edgemen must have two points of contact while working near the edge. The rescuers feet count as one point provided the rescuer does not intend to put body weight on the belay line and that the edge is not sloping.

An unweighted belay line is required if the edgemen intend to put body weight on their primary restraint.

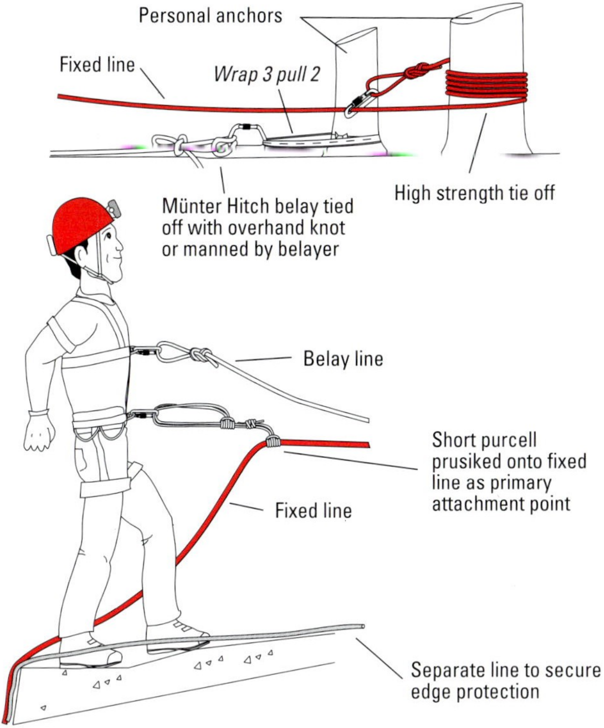

Personal anchors

Fixed line

Wrap 3 pull 2

High strength tie off

Münter Hitch belay tied off with overhand knot or manned by belayer

Belay line

Short purcell prusiked onto fixed line as primary attachment point

Fixed line

Separate line to secure edge protection

Edge Protection

- Do not put belay line in high directional
- Avoid standing under loaded working line
- Be certain to secure low directional to prevent losing it

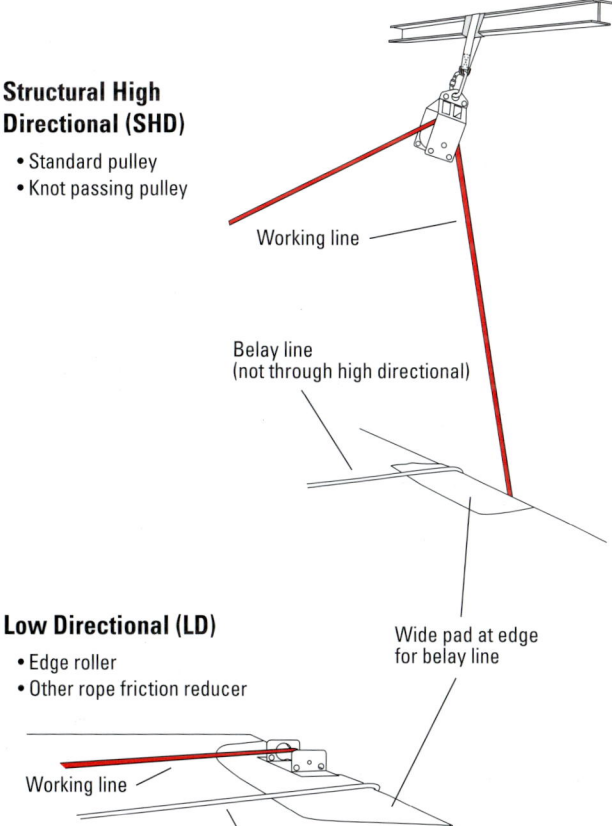

Structural High Directional (SHD)

- Standard pulley
- Knot passing pulley

Working line

Belay line
(not through high directional)

Low Directional (LD)

- Edge roller
- Other rope friction reducer

Working line

Wide pad at edge
for belay line

Belay line

Tandem Prusik Belay Setup

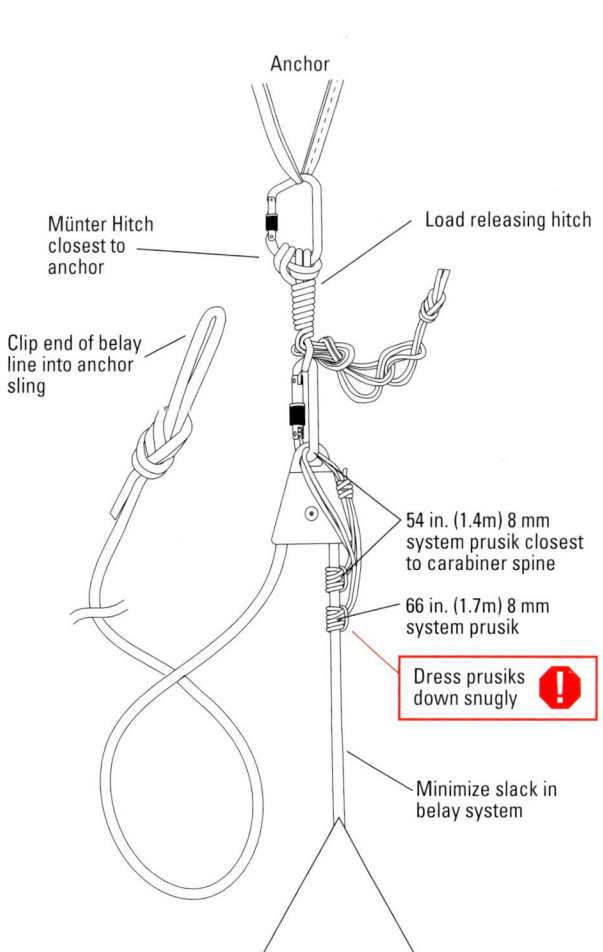

Anchor

Münter Hitch
closest to
anchor

Load releasing hitch

Clip end of belay
line into anchor
sling

54 in. (1.4m) 8 mm
system prusik closest
to carabiner spine

66 in. (1.7m) 8 mm
system prusik

Dress prusiks
down snugly

Minimize slack in
belay system

Rescue package

Tandem Prusik Belay Operation

Lower

1. For lowering system, place one hand on both prusiks and create a z-turn with the rope on the load side with the other hand.
2. Hold the z-turn and let rope out 1 to 2 ft. (~.5m) or 2/3 arm length.
3. Begin to turn z-turn hand while feeling and maintaining the tension on the rope.
4. Quickly move the z-turn hand back toward the prusik hand and make another z-turn.
5. Repeat this process always keeping a feel for the rescue package.

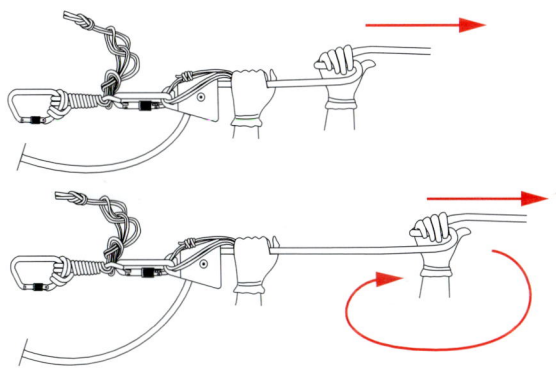

Raise

1. Pull constant tension on the free end of the belay rope.
2. Let the prusik minding pulley (PMP) mind the tandem prusiks.

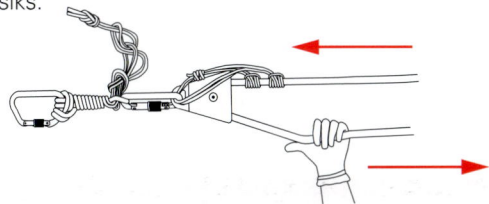

Technical Evacuation System

- Steep angle evacuation, raise and lower
- High angle evacuation, raise and lower
- Team based pick-off, raise and lower

Minimum setup requirements

Equipment
1 Working line kit w/rope
1 Belay line kit w/rope
1 Edge management kit
2 Extra ropes, 100-200 ft. (30-60m)
1 Patient immobilization device
1 Patient packaging kit
1 Bolt anchor kit (optional)

Personnel requirements
1 TSO
1 Safety
1 Belayman
1 Working line
1-4 Person litter team
1-2 Edgemen

Technical Sector Officer responsibilities for lowering or raising operation

1. Identify fall line.
2. Assign working line, belay line, litter crew, edgeman and safety officer.
3. Make rescue plan clear to all personnel.
4. Safety check all system components prior to pre-tensioning.
5. Pre-tension system within the safe zone.
6. Lower or raise rescue package only after final safety check is complete.

✓ The TSO has the final say!

Technical Evacuation System: Lower

Riggers (working and belay lines) role
- Get the plan from the TSO
- Gather required equipment
- Give the working and belay lines to the litter attendant
- Set up anchors
- Build system and prepare for operation

Edgemen role
- Find personal anchor
- Be prepared to rappel and ascend back up with belay
- Protect the working and belay lines from edge trauma
- Help the attendant negotiate the edge

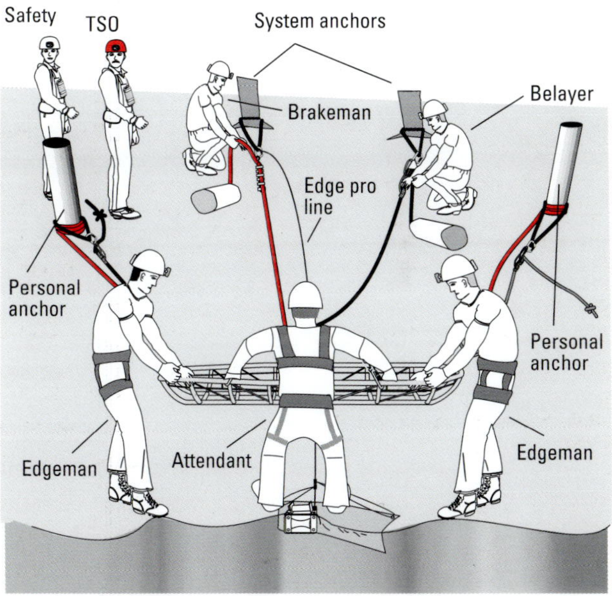

Safety TSO

System anchors

Brakeman

Belayer

Edge pro line

Personal anchor

Personal anchor

Edgeman   Attendant   Edgeman

✓ Pre-tension system away from edge prior to operation.

Technical Evacuation System: Raise

Mechanical advantage (MA) factors for raising

- Number of haulers
- Manned ratchet
- Change of direction (CD)
- Integral system or ganged system (see p. 41)
- Weight of load

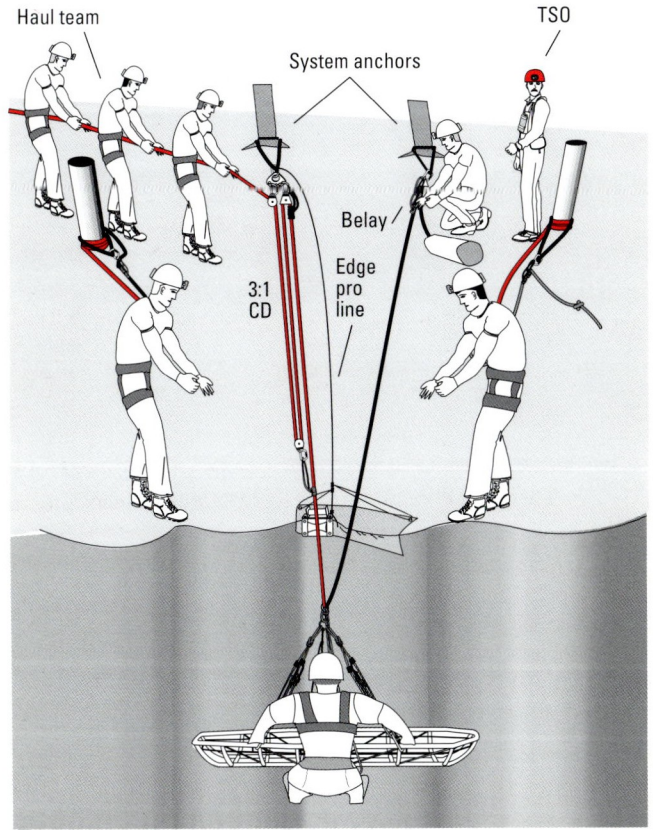

Haul team

TSO

System anchors

Belay

Edge
pro
line

3:1
CD

Technical Evacuation System Commands

TSO commands for lower

When all stations report rigging complete the TSO uses following procedure:

- ☐ Everyone stand by to pre-tension the system
- ☐ Safety is everything ok? Solve any problems
- ☐ Attendant or bearers ready? *Attendant ready*
- ☐ Edgemen ready? *Edgemen ready*
- ☐ On belay? *Belay on*
- ☐ Working line, take up rope, load & lock 6 bars*, advise when ready Working line loaded and locked 6 bars*
- ☐ Pre-tension the system Litter loads system
- ☐ System safety check Check entire system
- ☐ Safety checks, any problems? Solve any problems
- ☐ Attendant ready for lower? *Attendant ready*
- ☐ Edge ready? *Edge ready*
- ☐ Belay ready? *Belay ready*
- ☐ Brakeman ready? *Brakeman ready*
- ☐ Down slow Brakeman repeats *Down slow*

TSO commands for raise

- ☐ Safety is everything OK? Solve any problems
- ☐ Attendant or bearers ready? *Attendant ready*
- ☐ Edge control ready? *Edge ready*
- ☐ On belay? *Belay on*
- ☐ Haul team ready? *Haul team ready*
- ☐ Pre-tension the system Litter loads system
- ☐ System safety check Check entire system
- ☐ Safety checks, any problems? Solve any problems
- ☐ Attendant ready for raise? *Attendant ready*
- ☐ Edge ready? *Edge ready*
- ☐ Belay ready? *Belay ready*
- ☐ Haul team ready? *Haul team ready*
- ☐ Up slow Haul team repeats *Up slow*

TSO commands to convert from lower to raise

- ☐ Brakeman, lock off (tie off) Brakeman locked off
- ☐ Rig for raise *Rig for raise*
- ☐ Attendant, prepare for downward movement during transition
- ☐ Haul team advise when ready

*Number of bars may vary with the weight of the load.

Technical Evacuation System: Lower

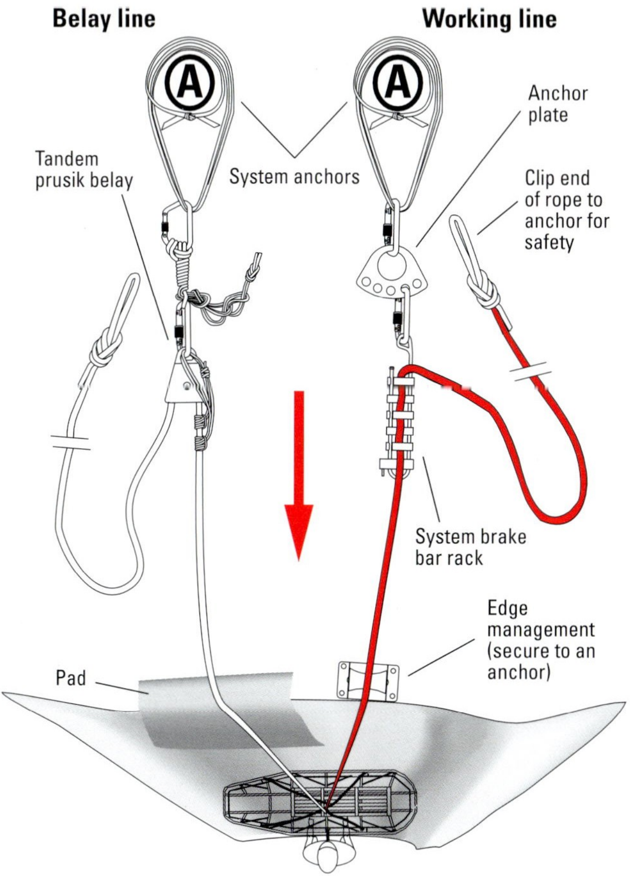

Belay line

Working line

Ⓐ System anchors Ⓐ

Anchor plate

Tandem prusik belay

Clip end of rope to anchor for safety

System brake bar rack

Edge management (secure to an anchor)

Pad

✓ Always designate a safety officer!

Technical Evacuation System: Raise

A ganged MA requires that a person operate the ratchet and keep slack out of the working line.

Construct mechanical advantage integral with working line or gang on a separate MA.

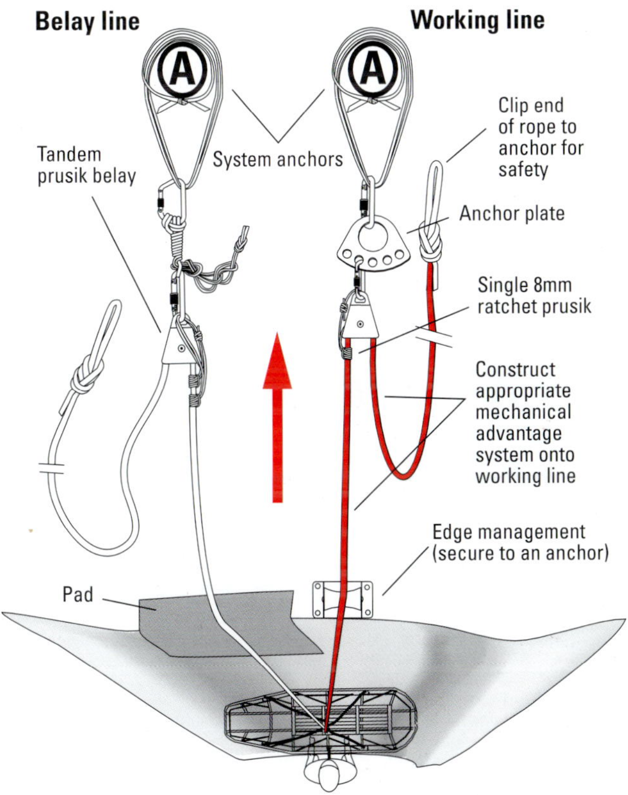

Belay line

Working line

Tandem prusik belay

System anchors

Clip end of rope to anchor for safety

Anchor plate

Single 8mm ratchet prusik

Construct appropriate mechanical advantage system onto working line

Edge management (secure to an anchor)

Pad

Steep Angle Evacuation 40° - 60°

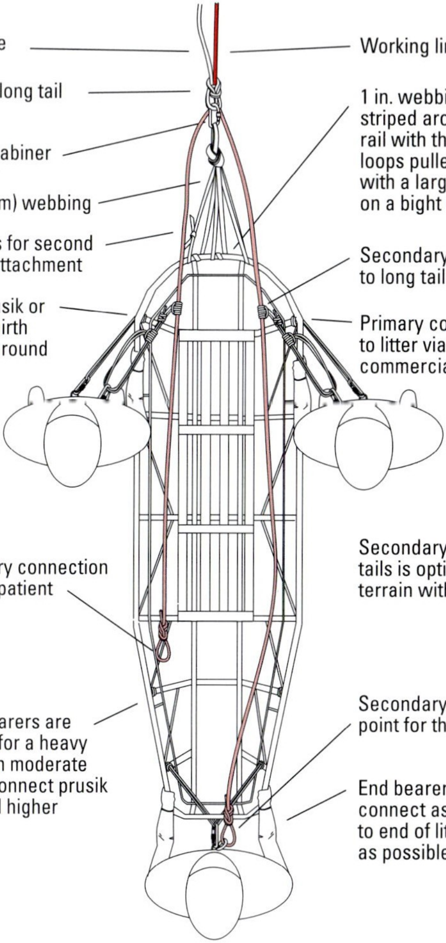

Belay line

Doubled long tail bowline

Steel carabiner or tri link

15 ft. (4.5m) webbing

Long tails for second point of attachment

Short prusik or purcell, girth hitched around strut

Secondary connection point for patient

If four bearers are required for a heavy patient on moderate terrain, connect prusik or purcell higher

Working line

1 in. webbing candy striped around litter rail with three separate loops pulled and tied off with a large overhand on a bight

Secondary connection to long tail via purcell

Primary connection to litter via prusik or commercial bearer strap

Secondary tie in to long tails is optional on easy terrain with no exposure

Secondary connection point for third bearer

End bearer must connect as close to end of litter as possible

✓ Caution: extreme forces can be generated during a steep angle operation. Choose anchors accordingly!

High Angle Litter Rigging

1. The litter attendant should receive the working line and belay line from the other riggers as soon as possible.

2. Tie a doubled long tail bowline through the litter bridle ring.

3. Leave 8 ft. (2.5m) of tail on each and tie a figure eight on a bight in each end.

4. Girth hitch a short purcell or commercial adjustable strap to the ring for the primary rescuer tie-in.

5. Tie one of the rope tails to the rescuer's harness as a second point of attachment. The second tail is for the patient or a second attendant.

6. The attendant should be prepared to clear vomitus from the patients airway by tilting the litter with a flip line or by using a hand powered suction unit.

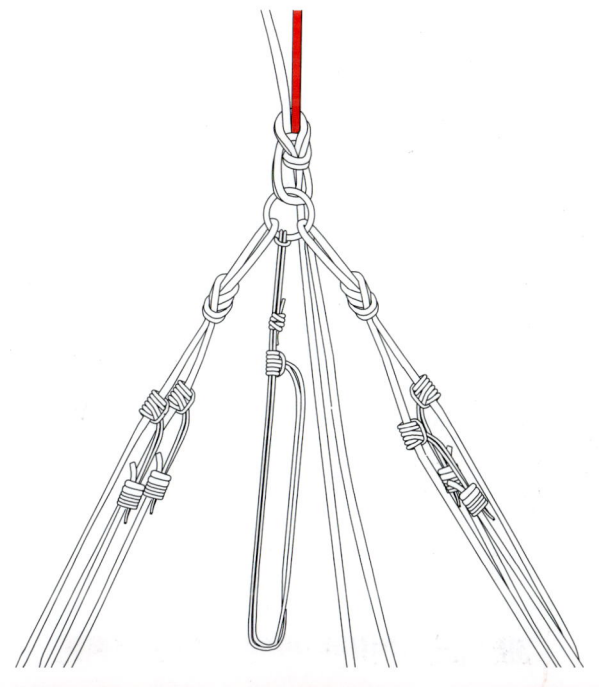

High Angle Evacuation 60° - 90°

- Consider suction device and barf line (used to tilt litter horizontally) to assist in clearing airway
- Remember packaging material
- The attendant controls the speed of the litter
- Go slow at the edge
- Be prepared to adjust the litter attitude
- Attach 15 ft. (4.5m) webbing to head and foot of litter for edgemen to assist litter attendant over the edge

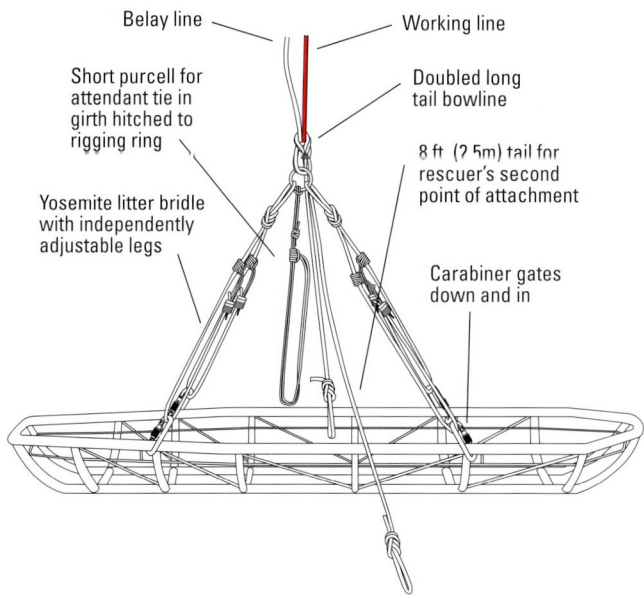

Belay line

Working line

Short purcell for attendant tie in girth hitched to rigging ring

Doubled long tail bowline

8 ft. (2.5m) tail for rescuer's second point of attachment

Yosemite litter bridle with independently adjustable legs

Carabiner gates down and in

Simple Mechanical Advantage Systems

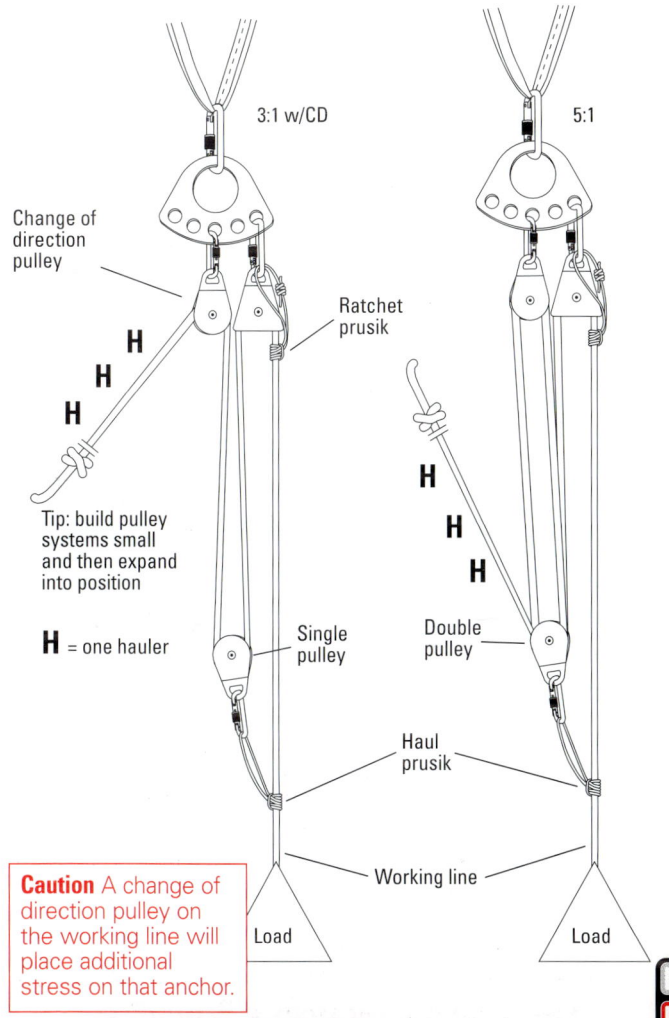

3:1 w/CD

5:1

Change of
direction
pulley

Ratchet
prusik

H

H

H

H

Tip: build pulley
systems small
and then expand
into position

H = one hauler

H

H

H

Single
pulley

Double
pulley

Haul
prusik

Working line

Caution A change of
direction pulley on
the working line will
place additional
stress on that anchor.

Load

Load

Compound Mechanical Advantage

A compound mechanical advantage is a simple MA pulling on the haul of another simple MA.

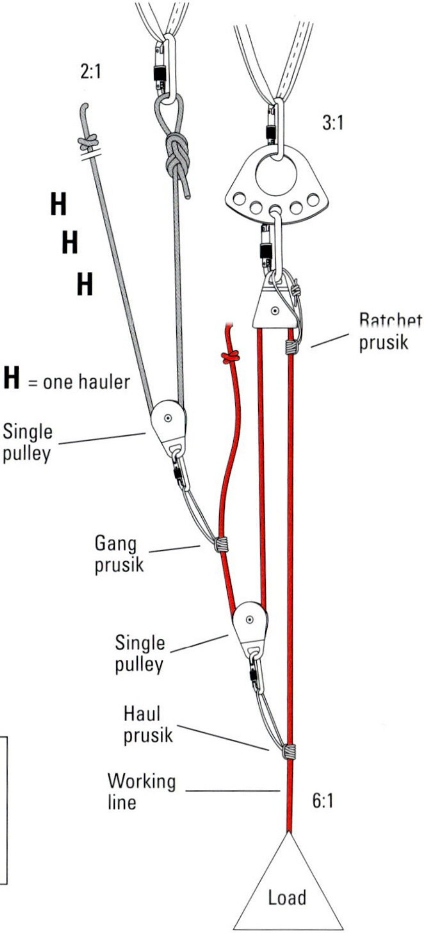

2:1

3:1

H
 H
  H

H = one hauler

Ratchet prusik

Single pulley

Gang prusik

Single pulley

Haul prusik

Working line

6:1

Load

Tip Offset the ganged MA to an anchor farther back to give the system more throw length.

Ganged Haul System

Use a ganged haul system when:

- There is not enough working line left to construct an integral mechanical advantage
- Time and resources allow riggers to construct an accessory pulley system which can be quickly ganged on to the working line at any time
- A knot must be passed on the working line

While using a ganged haul system:

- A ratchet is required on the working line
- A ratchet person must be assigned to take up the working line and mind the ratchet
- You do not need a ratchet on the pulley system

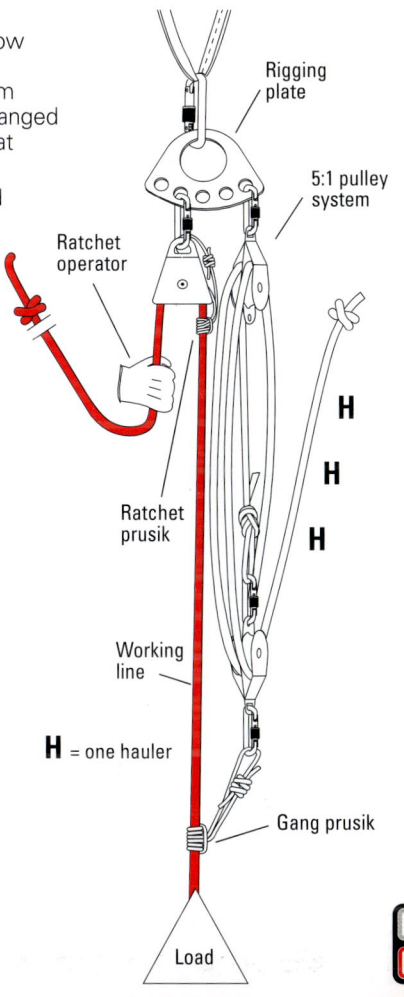

Rigging plate

5:1 pulley system

Ratchet operator

H

H

H

Ratchet prusik

Working line

H = one hauler

Gang prusik

Load

Conversion from Lower to Raise

1. Anticipate the need to switch under tension and be ready with a 6 ft. (2m) webbing loop, short system prusik and prusik minding pulley.
2. On command, brakeman locks off system brake rack (tie off optional) (Step 1).
3. Assistant attaches short system prusik to working line (Step 2).
4. Double over 6 ft. (2m) webbing loop and attach to anchor plate as spacer.
5. Attach PMP to working line between ratchet prusik and system brake rack.
6. Connect ratchet prusik and PMP to webbing loop (prusik closest to spine of carabiner).
7. Set ratchet prusik.
8. Slowly transfer load from system brake rack to ratchet prusik.
9. Remove system brake rack and construct raising system either integrally or ganged (Step 3).

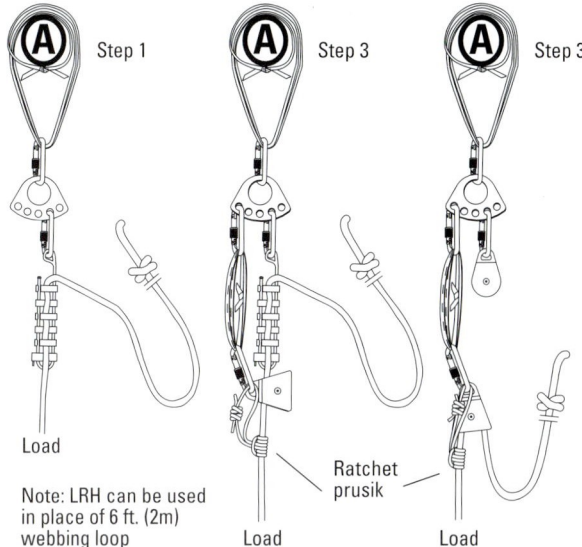

Step 1

Step 3

Step 3

Load

Note: LRH can be used in place of 6 ft. (2m) webbing loop

Ratchet prusik

Load

Load

Knot Passing

Be prepared to pass a knot whenever you suspect that your ropes do not reach the ground.

A knot pass on the working line during lowering requires:

- One LRH
- 8mm system prusik
- An assistant

1. Tie the new rope onto the starting rope with a double fishermans bend well ahead of time.
2. Have the assistant connect the LRH to the rigging plate (step 1).
3. Apply a three-wrap prusik below the brake bar rack and while minding it, attach it to the LRH.
4. Set the prusik when the knot is 12 in. (30cm) from the brake bar rack (step 2).
5. Remove the rope from the break bar rack and reinstall it with the knot below the brake bar rack.
6. When the brake bar rack is locked off, release the LRH and transfer the load back to the brake bar rack (step 3).
7. Remove the prusik and continue lowering.

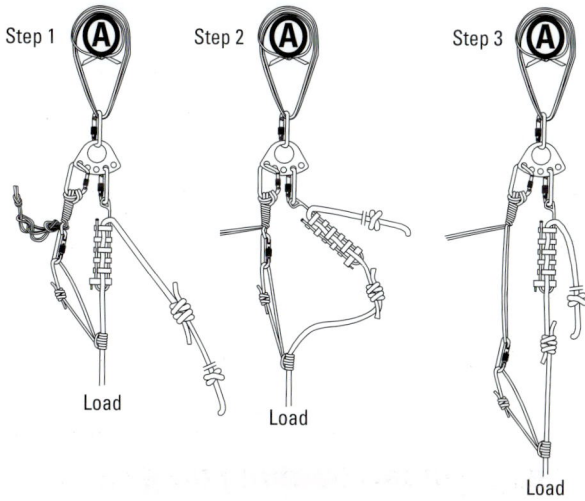

Step 1   Step 2   Step 3

Load   Load   Load

Knot Passing: Lower

A knot pass on the belay line during lowering requires:
- One additional tandem prusik belay set-up
- An assistant

1. Have assistant install duplicate belay setup on the other side of the knot (same procedure for lower or raise).
2. When the knot approaches the belay, clip the duplicate belay into the lower bight of the LRH next to the primary belay.
3. Belayer switches to the new belay and the assistant removes the old belay.
4. Warning! Do not disconnect the original belay setup until the new one is in operation.

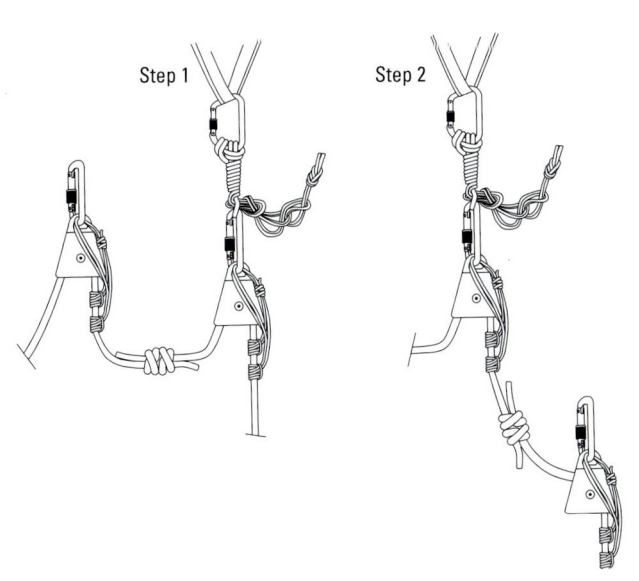

Step 1

Step 2

Knot Passing: Raise

Passing a knot in the working line on raise

1. Use a ganged mechanical advantage for your haul system.
2. When the knot approaches, add a new gang prusik below the knot and move the pulley system to the new gang prusik.
3. Duplicate the ratchet below the knot.
4. Clip the new ratchet onto the anchor when the knot reaches the old ratchet.
5. Continue with the raising.

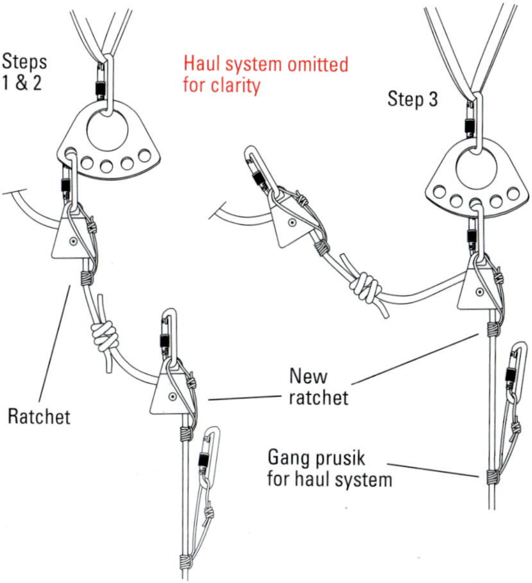

Steps
1 & 2

Haul system omitted
for clarity

Step 3

New
ratchet

Ratchet

Gang prusik
for haul system

Mid-Face Litter Scoop

1. Use when injured person is hanging on rope or a fall arrest system in mid rock face or mid structure.

2. Send rescuer down to patient on rappel for assessment and basic stabilization while litter is being rigged.

3. Pre-rig litter bridle with rescue mini 4:1 CD from rigging ring to foot of litter using a 33 ft. (10m) 9mm cordalette.

4. Leave carabiners on foot legs of bridle unlocked.

5. Lower down to a point 10 ft. (3m) above patient and stop.

6. Disconnect foot bridle section and lower litter into scoop position.

7. Slowly lower until even with patient and stop.

8. Stop early, it is easier to lower to correct position than to raise.

9. Begin working patient into litter, raise foot of litter gradually.

10. Move patient a little at a time until litter is level.

11. Re-connect foot of litter bridle and secure patient into litter.

12. Lower or raise as needed.

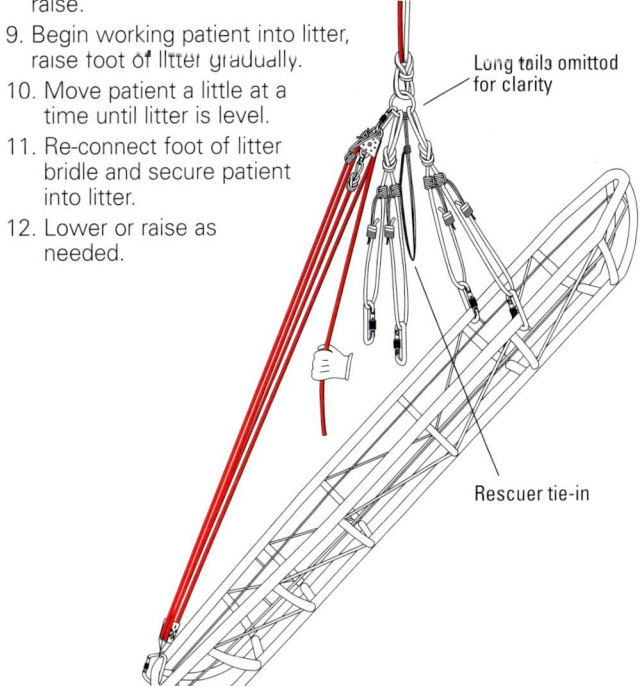

Long tails omitted for clarity

Rescuer tie-in

Mid-Face Litter Scoop

Rigging

1. Tie working line and belay line to yoke ring with doubled long tail bowline.
2. Girth hitch short purcell to ring for attendant primary attachment.
3. Connect one of the rope tails to attendant for second point of attachment.
4. Connect top of mini-rescue 4:1 CD to rigging ring where ratchet prusik is in reach of rescuer.
5. Extend MA and attach bottom to foot of litter.
6. Leave foot bridle carabiners unlocked to facilitate easy disconnect.
7. Secure any patient packaging supplies.

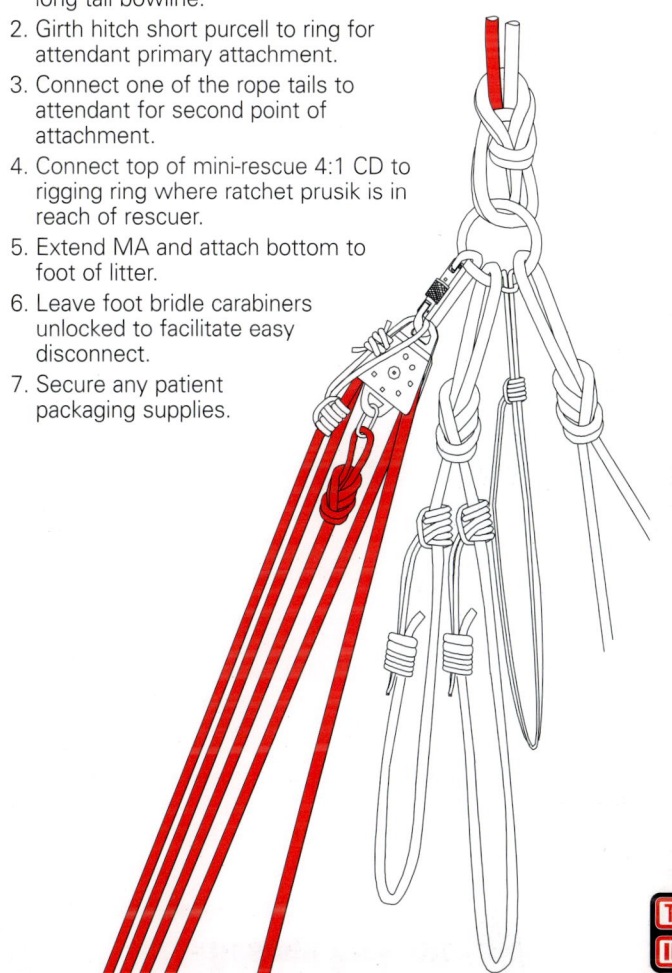

Rescue Pick-off

Non-injured or slightly injured subject:

Unsupported

- Subject is standing or clinging from rock or structure without a rope or fall arrest system.
- Be prepared to capture the subject upon contact.

Supported

- Subject is hanging on a rope or fall arrest system.
- Subject is wearing some type of harness.
- Rescuer must transfer subject's weight from subject's system to rescue system.
- Consider suspension trauma (pg. 152)

Unsupported   **Supported**

Rescue Pick-off Types

Rescuer based

- Single rescuer rappels to subject using 6 bar brake bar rack and tandem prusik belay

Minimums

2 Technicians
1 Working line kit
2 Life safety ropes
1 Rescue harness
1 Mini rescue 4:1CD
1 Pick-off strap
1 Spare helmet

Pros

- Fastest access to a subject in a remote spot
- Requires a minimum of technicians and gear

Team based

- Team lowers a rescuer to the subject with a working line and belay line with a technical evacuation system

Minimums

4 Technicians
1 Working line kit
1 Belay line kit
2 Life safety ropes
1 Rescue harness
1 Pick-off strap
1 Spare helmet

Pros

- Rescuer has hands free and better able to deal with the subject
- Easy to bring rescue package back up to the top
- Provides the most options in the event of a problem

Rescuer Based Pick-off: Unsupported

Subject unsupported setup

1. Locate fall line above subject and set up slightly right or left if possible.
2. Establish anchor system for rescuer rappel.
3. Establish tandem prusik belay.
4. Pre-connect rescue harness to brake bar rack and belay.
5. Replace rescue harness into bag and attach to gear loop.
6. Bring helmet for subject, extra webbing and carabiners.
7. Safety check with belayer and carefully rappel to subject. (continued p. 52)

Warning!
Watch out for side loaded carabiners.

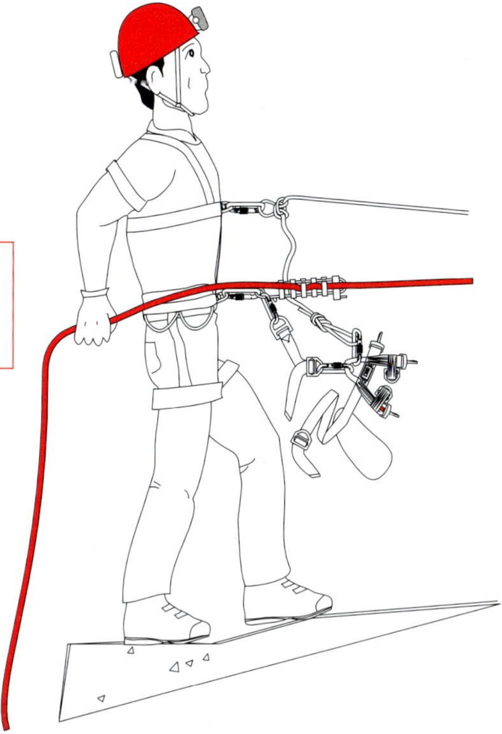

Rescuer Based Pick-off Setup

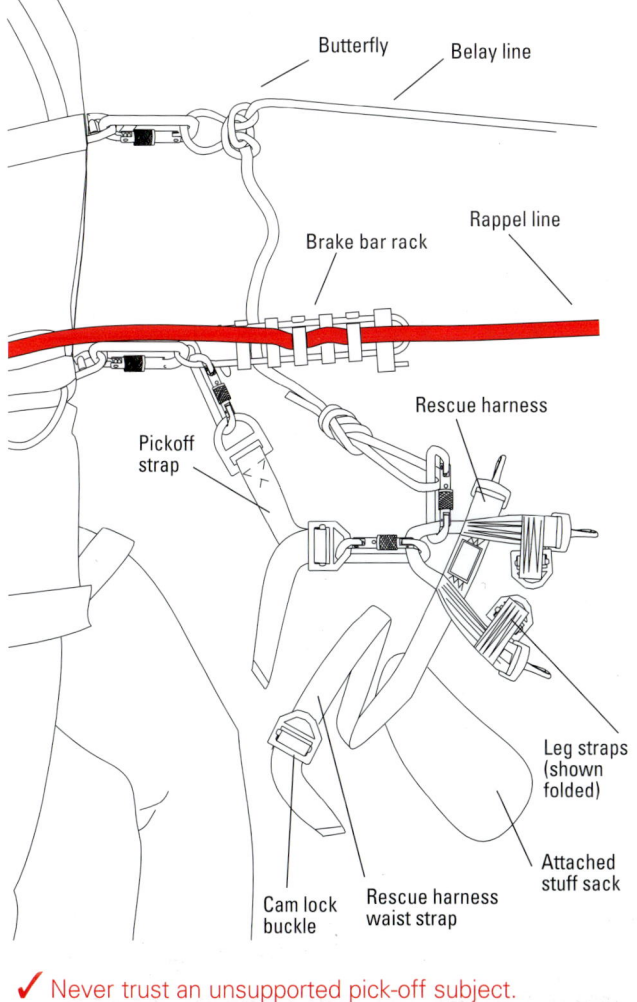

Butterfly

Belay line

Rappel line

Brake bar rack

Rescue harness

Pickoff
strap

Leg straps
(shown
folded)

Attached
stuff sack

Cam lock
buckle

Rescue harness
waist strap

✓ Never trust an unsupported pick-off subject.

Rescuer Based Pick-off: Unsupported

Subject unsupported setup

8. Approach subject with caution, reassure and instruct them not to move or grab the rope.

9. Stop slightly above the subject.

10. Add all bars, lock off and tie off brake bar rack.

11. Remove rescue harness from pouch and orient it with the label toward the subject's navel.

12. Connect yellow waist belt around subjects waist and cinch tight (be careful not to pull subject off).

13. Connect each leg loop.

14. Put helmet on subject.

15. Apply chest harness if needed.

16. Instruct subject to sit back onto harness.

17. Untie brake bar rack and rappel to ground with subject.

18. Remove subject from fall line ASAP.

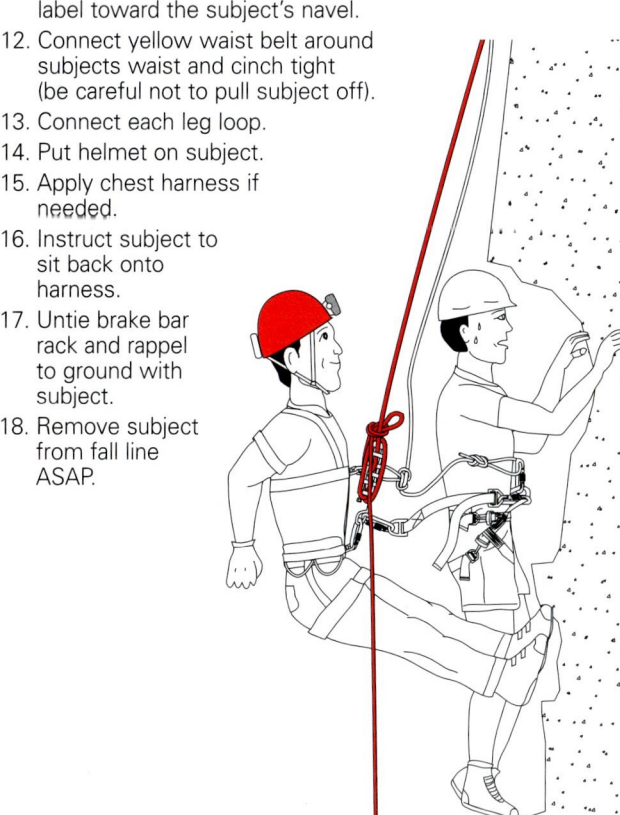

Rescuer Based Pick-off: Supported

Subject supported setup

1. Use same rappel setup as unsupported except not using rescue harness.
2. Bring mini rescue 4:1CD mechanical advantage down to subject (stowed in carry sack).
3. Bring helmet for subject.
4. Carefully rappel to subject and establish rapport as you approach.
5. Stop about 1 ft. (.3m) above subject.
6. Add all bars to brake bar rack and tie off.
7. Connect pick-off strap and belay line to subject's harness.
8. Place helmet on subject.
9. Apply chest harness to subject if possible.
10. Put short prusik on subject's support line. (continued)

Make every effort to stop above the subject. Stopping below the subject makes the weight transfer very difficult.

Rescuer Based Pick-off: Supported

Subject supported setup

11. Connect mini rescue 4:1 CD to prusik on subject's rope.
12. Connect other end of mini rescue 4:1 CD to subject's harness.
13. Raise subject up until it is possible to disconnect subject from their system.
14. Disconnect the subject from their system.
15. Release mini rescue 4:1 CD and lower subject onto pick-off strap.
16. Disconnect pick-off mini rescue 4:1 CD from subject harness and from the prusik.
17. Remove prusik and stow both in pouch.
18. Explain descent method to subject.
19. Untie brake bar rack and descend.

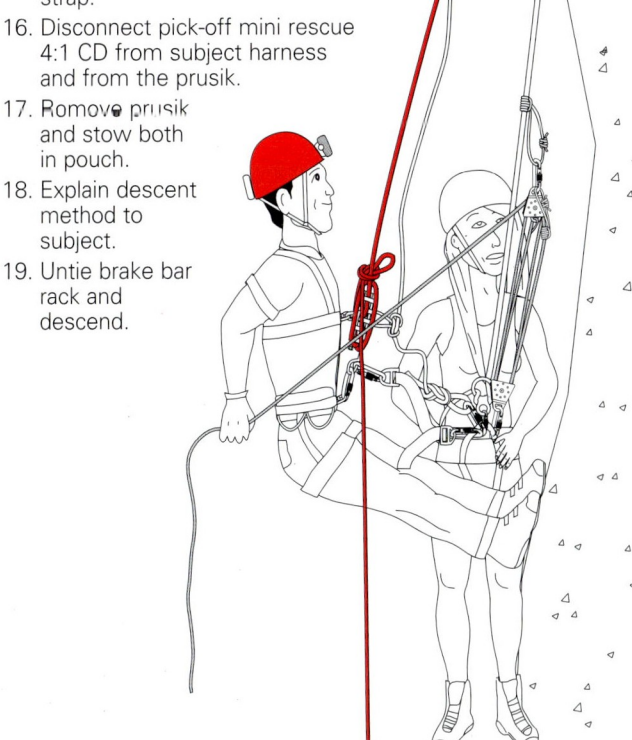

Team Based Pick-off: Unsupported

Subject unsupported setup

1. Rescuer connects to working line and belay line that are tied together as one with doubled long tail bowline and 3 ft. (1m) tails. Figure 8 on a bight is put in each tail.
2. Connect one tail to rescuer as second point of contact.
3. Connect other tail to rescue harness.
4. Attach long purcell to working line and connect adjustable loop to rescue harness as primary attachment point.
5. Bring helmet for subject.
6. Advise TSO that rescuer is ready for lower.
7. Be sure to protect ropes from edge friction. (continued)

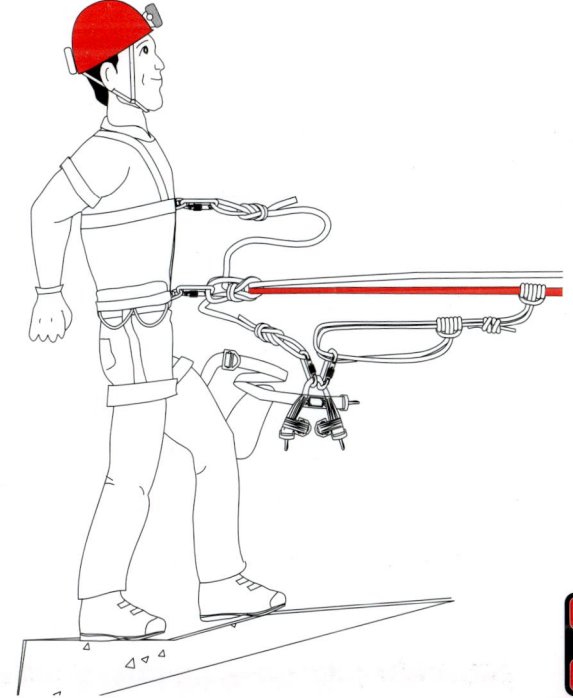

Team Based Pick-off: Unsupported

Subject unsupported setup

8. Lower over edge when ready.

9. Have the pre-connected pick-off harness in hand ready to capture the subject as soon as you are in position.

10. Call for a stop when even with the subject.

11. Carefully attach waist strap around subject with label in toward subject's navel.

12. Attach leg loops.

13. Adjust purcell up so that it is tensioned.

14. Put helmet on subject.

15. Put chest harness on subject if needed.

16. Advise TSO of intentions to go up or down.

17. Have subject put weight onto harness.

18. Advise TSO when ready to raise or lower.

Team Based Pick-off: Supported

Subject supported setup

1. Setup is the same as unsupported, except that it does not use rescue harness.
2. Call for stop slightly above subject and establish rapport.
3. Advise TSO to rig for raise.
4. Connect purcell to subject's harness (back up with webbing if necessary).
5. Adjust tension on purcell.
6. Connect belay to subject's harness.
7. Put helmet on subject.
8. Put chest harness on subject if needed.
9. Request raise to unload subject's support system.
10. Disconnect subject's support system.
11. Advise TSO when ready for raise or lower.

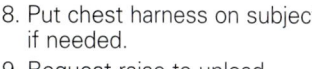

Confined Space Command Checklist Ⓐ

Phase I: Size-up
☐ Primary assessment
- ☐ Secure witness or competent person
- ☐ Identify immediate hazards
- ☐ Location, number, condition of victims
- ☐ Attempt contact
- ☐ Secure entry permit

☐ Secondary assessment
- ☐ What type of space
- ☐ Products in space
- ☐ Hazards: atmospheric, mechanical, electrical
- ☐ Diagram of space
- ☐ Can non-entry retrieval be made?
- ☐ Structural stability of space
- ☐ Proper personnel and equipment on scene
- ☐ Additional resources necessary
- ☐ Atmospheric monitoring: ventilation, respiratory, retrieval system
- ☐ Rescue or recovery/survivability profile

Phase II: Pre-entry operations
☐ Initiate Fire Department confined space rescue permit

☐ Make general area safe
- ☐ Establish perimeter
- ☐ Evacuate if necessary
- ☐ Traffic and crowd control

☐ Make rescue area safe
- ☐ Establish lobby control accountability
- ☐ Test atmosphere: oxygen, flammable, toxic
- ☐ Ventilate appropriately for space
- ☐ Secure hazards: lock-out, tag-out

☐ Action plan with back-up plan

☐ Entry team ready
- ☐ Back-up team in place

Confined Space Command Checklist

- ☐ Proper equipment
    - ☐ Personal protective equipment
    - ☐ Explosion proof lighting and communications
    - ☐ Respiratory system (SCBA, SABA)
    - ☐ Remote air monitoring
    - ☐ Personal atmospheric monitor
    - ☐ Class 3 harness
- ☐ Rigging team
    - ☐ Tripod, davit or crane
    - ☐ Retrieval system with back-up system
    - ☐ Patient packaging devices
- ☐ Air supply
    - ☐ Primary air supply
    - ☐ Back-up team air supply
    - ☐ Utility truck high pressure hook-up for refill of bottles
- ☐ Pre-entry briefing
    - ☐ Advise each team of expected task
    - ☐ Discuss emergency procedures for each team
    - ☐ Provide each team with site briefing
    - ☐ Advise each team of time limits

Phase III: Entry and rescue operations

- ☐ Entry system safety check
- ☐ Make entry
    - ☐ Continual atmospheric monitoring
    - ☐ Constant communication with the entry team
    - ☐ Monitor ventilation system
    - ☐ Assist entry team with line management
- ☐ Locate victim
    - ☐ Patient packaging and extrication

Phase IV: Termination

- ☐ Personnel accountability report
- ☐ Remove tools and equipment
- ☐ Decontamination
- ☐ Secure scene
- ☐ Consider debriefing
- ☐ Call OSHA

A

Confined Space Rescue

OSHA 29 CFR 1910.146 applies to general industry and the rescue service.

An OSHA confined space is defined as:

1. A space large enough for personnel to physically enter.
2. Not designed for continuous employee occupancy.
3. An area with limited entry and egress.

A confined space permit is required if the space has one or more of the following hazards:

1. Atmospheric hazards.
2. Configuration hazard.
3. Engulfment hazard.
4. Any other recognized hazard.

Acceptable entry conditions

Oxygen between: 19.5% and 23.5%
Lower explosive limit (LEL):   <10% of the LEL
Toxicity: < IDLH

Immediately dangerous to life and health (IDLH)

! Heat stress can quickly become a life threatening hazard. Rotate crews frequently.

✓ Take the extra time to carefully manage all lines.

✓ Be sure to have the lobby/attendant take up all lines as the entry team returns to the outside.

✓ Expect the atmosphere to suddenly become unsafe.

✓ Monitor the atmosphere continuously.

Confined Space Entry Safety Checklist

TSO commands for confined space entry

- ☐ Everyone take positions and prepare for entry checklist
- ☐ Attendant ready? *Attendant Ready*
- ☐ Retrieval ready? *Retrieval ready*
- ☐ Entry team egress bottle pressures? *Record pressure*
- ☐ Backup team egress bottle pressures? *Record pressure*
- ☐ Personal air monitor on *Monitor checks ok*
- ☐ Mechanical ventilation on *Ventilation on*
- ☐ Air supply ready? *Air supply ready*
- ☐ Entry team go on air *On air*
- ☐ Primary comm check *Primary comm OK*
- ☐ Secondary comm check *Secondary OK*
- ☐ Primary light check *Primary light OK*
- ☐ Secondary light check *Secondary OK*

- ☐ Entry team ready? *Entry team ready*
- ☐ Backup team ready? *Backup team ready*

- ☐ System safety check, any problems? Solve any problems
- ☐ Entry team make entry *Making entry*
- ☐ Attendant note time of entry *Time noted*

Emergency checklist for backup/rescue team entry

- ☐ Attendant ready? *Attendant ready*
- ☐ Retrieval ready? *Retrieval ready*
- ☐ Backup air supply ready? *Air supply ready*
- ☐ Rescue team go on air *On air*
- ☐ Personal air monitor on *Monitor checks ok*
- ☐ Primary comm check *Primary comm OK*
- ☐ Secondary comm check *Secondary OK*
- ☐ Primary light check *Primary light OK*
- ☐ Secondary light check *Secondary light OK*

- ☐ Rescue team ready? *Rescue team ready*

- ☐ Safety checks, any problems? Solve any problems

- ☐ Rescue team make entry *Making entry*
- ☐ Attendant note time of entry *Time noted*

Personal Protective Equipment

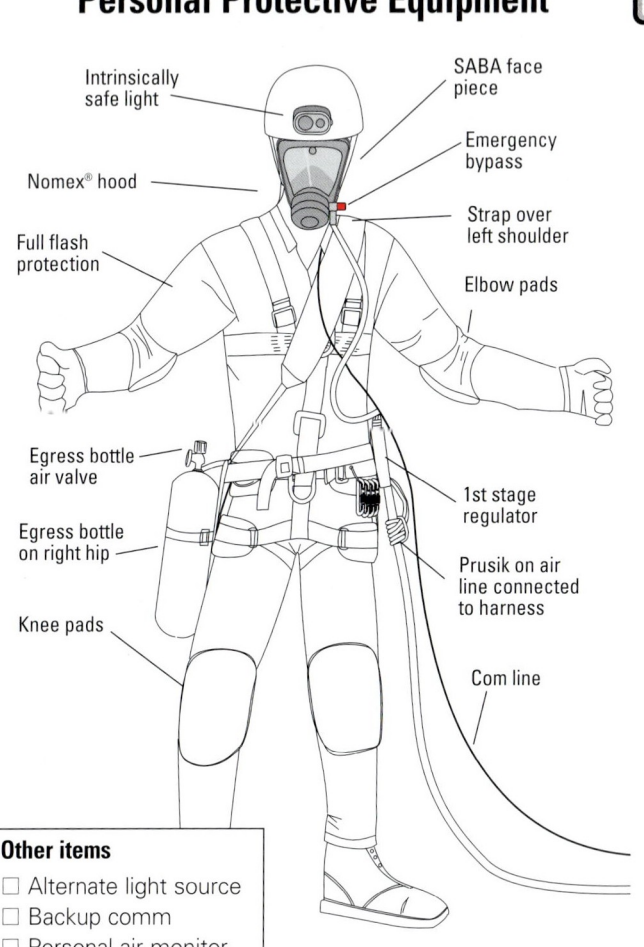

Intrinsically safe light

SABA face piece

Emergency bypass

Nomex® hood

Strap over left shoulder

Full flash protection

Elbow pads

Egress bottle air valve

1st stage regulator

Egress bottle on right hip

Prusik on air line connected to harness

Knee pads

Com line

Other items

☐ Alternate light source
☐ Backup comm
☐ Personal air monitor

A confined space harness must have a dorsal connection ring.

✓ Egress time should not exceed egress bottle capacity!

Supplied Air Station Operation

Setup

1. Assemble required equipment:
   - Remote air carts
   - SABA
   - Up to 300 ft. (90m) of hose per rescuer
   - Extra air bottles
2. Position air carts in close proximity to entry portal.
3. Stretch out all air hoses and unkink.
4. Connect hose to entry team connection port on air cart.
5. Stack hose in figure eight coils or long loops as each section is connected.
6. Consider tagging each section of hose to identify rescuer.
7. Connect hose to rescuer SABA.

Air cart operation (air cylinders only)

1. Confirm that both air tanks are full and that respirator regulator is turned fully counterclockwise.
2. Slowly turn on one of the two cylinders (primary). Alarm should sound briefly at initial start-up.
3. Confirm that outlet gauge pressure is set between 60 and 120 psi. without respirators attached (pressure will vary between different models of air supply cart).
4. When alarm sounds, open valve of secondary tank.
5. Alarm must stop before proceeding.
6. Close valve of used primary tank and replace with full tank.

SABA operation

1. Confirm that egress bottle is full.
2. Don SABA.
3. Keep egress bottle valve within easy reach at all times.
4. Connect air hose to first stage regulator.
5. Apply prusik to air hose and connect to harness.
6. Don face piece and test seal.
7. Don Nomex® hood and helmet.
8. Connect second stage regulator to mask.
9. Confirm operation of emergency bypass feature.

Remote Air Supply Cart

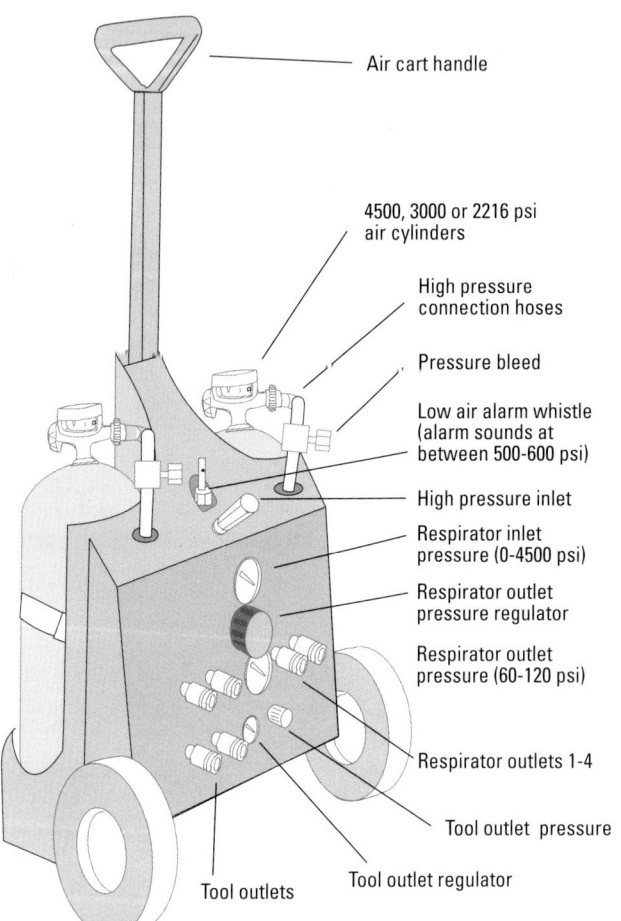

Air cart handle

4500, 3000 or 2216 psi
air cylinders

High pressure
connection hoses

Pressure bleed

Low air alarm whistle
(alarm sounds at
between 500-600 psi)

High pressure inlet

Respirator inlet
pressure (0-4500 psi)

Respirator outlet
pressure regulator

Respirator outlet
pressure (60-120 psi)

Respirator outlets 1-4

Tool outlet pressure

Tool outlet regulator

Tool outlets

Communication Position (Attendant)

Setup

1. Position intercom kit in close proximity to entry portal (watch for hazardous atmosphere near the portal).
2. Stretch out comm line and un-kink.
3. Connect required number of comm line sections together.
4. If connectors will not lock into position, clean O ring mating surfaces with moist rag.
5. Connect female end of comm line to command module.
6. Stack comm line so as to inhibit kinking problems.
7. Connect operator head set to operator connection port on command module.
8. Determine whether entry team rescuer will use headset or ear piece and throat mike.
9. Connect male end of comm line to rescuer.
10. Secure comm line to rescuer harness with small loop of slack between harness and end connection.
11. Install batteries in command module and test.

Operation

1. The attendant is required to maintain constant communication with the entry team.
2. The attendant can relay information to the TSO.
3. The TSO should not wear the headset unless it is a single side headset.
4. Adjust volume controls, as necessary.

Backup plan

1. The backup team must have a dedicated communication and air system.
2. Repeated contacts with entry team should be made via radio.
3. Test radio at junction points.
4. If communications fail, attempt brief troubleshooting and whistle or air horn signals (one long blast, repeat if necessary).
5. If communications cannot be re-established within one minute, send in the backup team.

Intrinsically Safe Intercom System

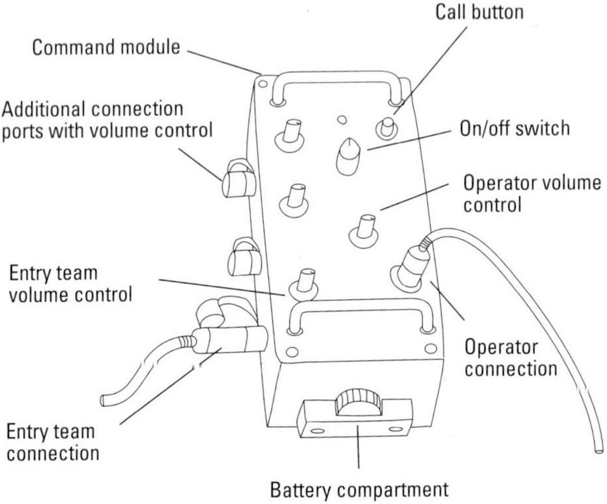

Call button

Command module

Additional connection
ports with volume control

On/off switch

Operator volume
control

Entry team
volume control

Operator
connection

Entry team
connection

Battery compartment

Have back-up batteries on hand!

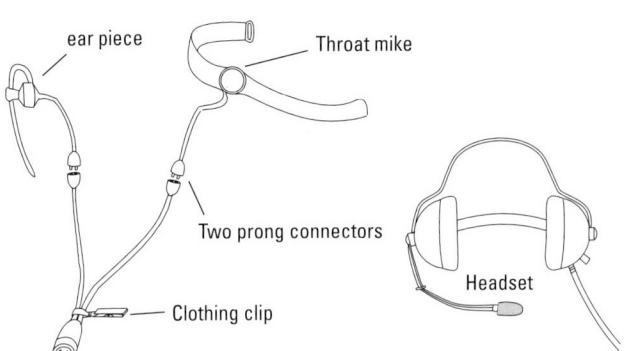

ear piece

Throat mike

Two prong connectors

Headset

Clothing clip

Atmospheric Monitoring

Principles of air monitoring

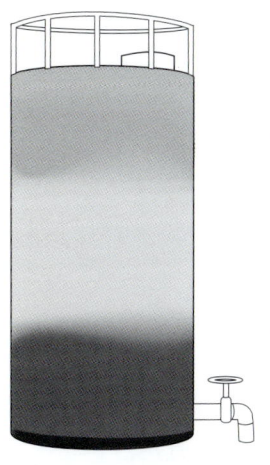

- Calibrate and span meter according to department procedures.
- If oxygen level is not normal, flammability readings will be affected.
- Spaces may have stratified atmospheres, all levels of space must be monitored.
- Allow for air intake in sampling hose at approximately 1 second per foot of hose.
- 10,000 parts per million = 1%.
- If oxygen reading is 1% low and it is being displaced by a contaminant, up to 5% of the total atmosphere may consist of that contaminant (50,000 ppm).
- Physical properties of a product can be found in the NIOSH pocket guide or MSDS.
- The calculated molecular weight of air is 29.

Below is an example of estimating the flammability and toxicity in a space in order to develop a victim survivability profile. A meter may not be required if the physical properties of the product are known.

Toluene Physical Properties		70°F (21°C) day	
Flash point	40°F (4.5°C)	>FP	yes
Molecular weight	92	>29	yes
LEL	1.1%	≥	yes
UEL	7.1%	≥	unknown
IDLH	500 ppm	est. ppm	11,000 ppm

✓ How long has the victim been down?
✓ Is this a body recovery?

Ventilation

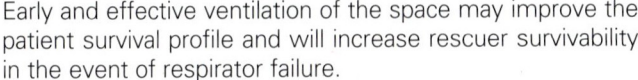

Early and effective ventilation of the space may improve the patient survival profile and will increase rescuer survivability in the event of respirator failure.

The capacity of the fan in cubic ft. per minute (CFM) divided into the volume of the space in cubic feet equals the time it takes to exchange the air one time.

Intrinsically safe axial fan

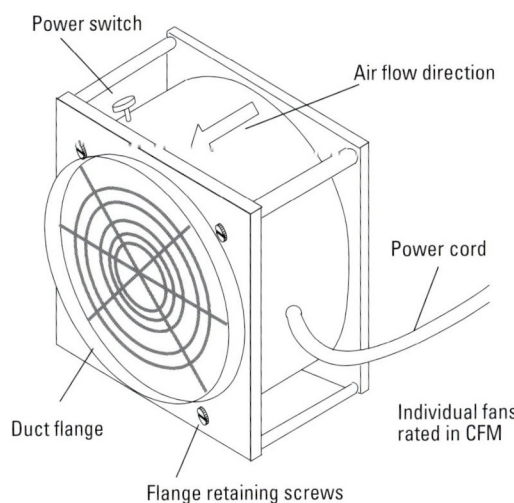

Power switch

Air flow direction

Power cord

Duct flange

Flange retaining screws

Individual fans rated in CFM

✓ The ventilation goal is to exchange the air in the space as many times as possible.

✓ Fan should be rated intrinsically safe and grounded.

✓ Place fans where they will have maximum effect, as close to the hazard as possible, but outside the contaminated area.

Ventilation System Components

The axial fan is capable of positive and negative ventilation depending upon which side the duct is connected. The fan shown is only able to exhaust with the 16 in. duct. A soft reducer coupling is not suitable for exhaust ventilation. Know your equipment.

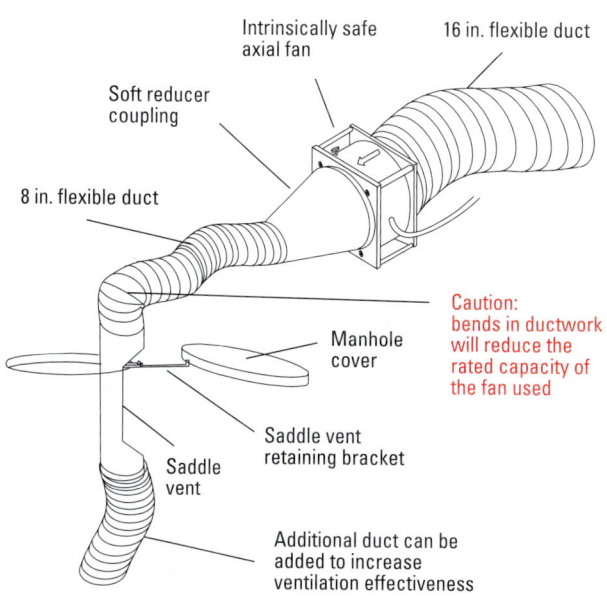

Intrinsically safe axial fan

16 in. flexible duct

Soft reducer coupling

8 in. flexible duct

Caution: bends in ductwork will reduce the rated capacity of the fan used

Manhole cover

Saddle vent retaining bracket

Saddle vent

Additional duct can be added to increase ventilation effectiveness

✓ Be extremely cautious when ventilating spaces with known flammable atmospheres due to the potential of the exhaust component reaching an ignition source.

Confined Space Ventilation

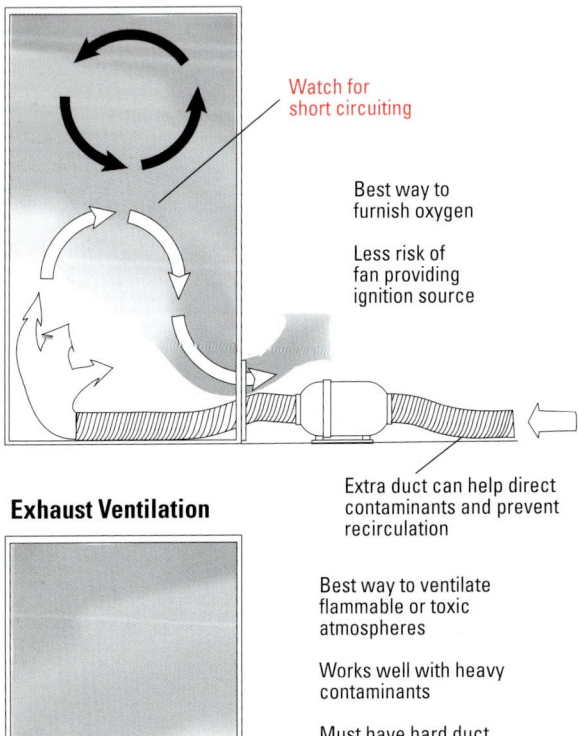

Supply Ventilation

Watch for
short circuiting

Best way to
furnish oxygen

Less risk of
fan providing
ignition source

Extra duct can help direct
contaminants and prevent
recirculation

Exhaust Ventilation

Best way to ventilate
flammable or toxic
atmospheres

Works well with heavy
contaminants

Must have hard duct

Watch for recirculation

Confined Space Ventilation

Look for other openings to make ventilation more effective

Consider whether the contaminant is heavy or light and set up ventilation accordingly

Supply Ventilation

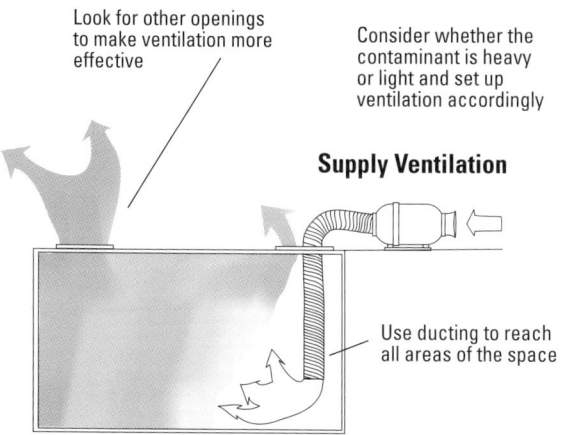

Use ducting to reach all areas of the space

Always consider where the contaminated exhaust is going and if it will pose an additional hazard

Supply/Exhaust Ventilation

Combination supply/exhaust ventilation is most effective

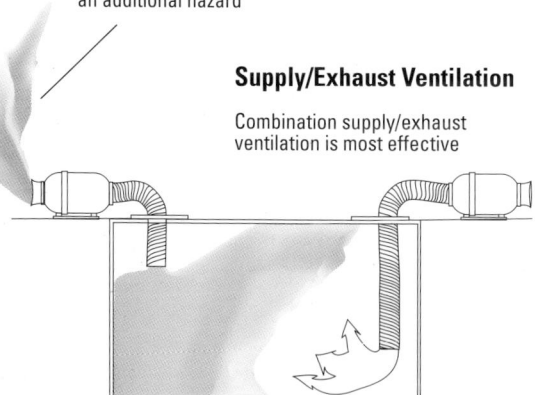

Extrication Device

Application

1. Apply cervical collar and maintain spinal stabilization.
2. Position extrication device behind patient and under arms.
3. Release strap retaining flaps one at a time as each strap is applied.
4. Loosely fasten chest straps.
5. Fasten shoulder straps.
6. Fasten groin straps.
7. Have patient take a breath and tighten chest straps.
8. Tighten shoulder and groin straps.
9. Fasten forehead and chin straps.
10. Connect lifting strap to lifting points.

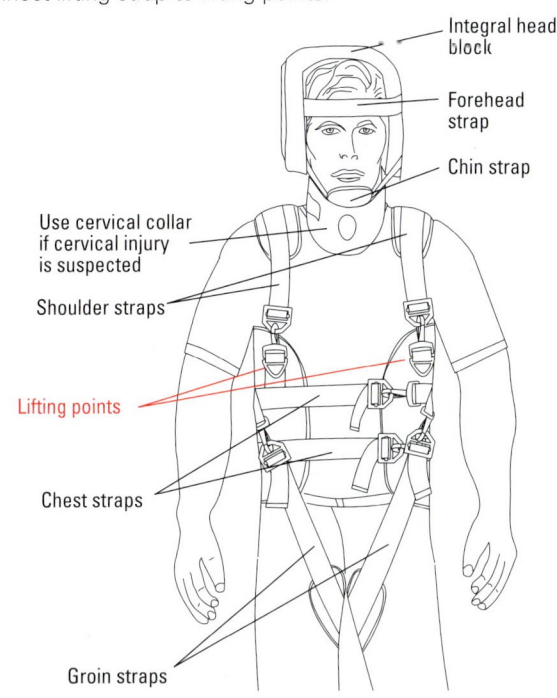

Integral head block

Forehead strap

Chin strap

Use cervical collar if cervical injury is suspected

Shoulder straps

Lifting points

Chest straps

Groin straps

Rescue Tripod and Winch

Setup

1. Remove the tripod from its carry case and stand upright.
2. Move each leg outward into the working position.
3. Slide legs up into the headpiece to engage the leg locks.
4. Remove the winch from its case.
5. Position the winch onto the fixed pin on the tripod leg mounting bracket.
6. Tilt the winch against the mounting bracket and insert the detente pin into the detente pin hole.
7. Place the crank handle on the low speed shaft.
8. With an assistant, reverse the winch and reel out approximately 8 ft. (2.5m) of cable.
9. Remove the cable retaining pins from the headpiece.
10. Place the cable over both guide wheels.
11. Replace the cable retaining pins.
12. Lower the counterweight until it is near ground level.
13. With three rescuers, adjust the height of each leg and install the leg adjustment pins.
14. Install and adjust the leg anti-spread chain.
15. Position the tripod over the opening.

Operation

1. Attach the crank handle to either the low-speed or high-speed shaft.
2. To raise, crank handle in the direction that it will move.
3. To lower, slightly raise and simultaneously apply downward pressure to the ratchet brake lever.
4. Lower cable while holding the ratchet brake lever in the down position.

! A rescue tripod is free standing and can easily collapse or tip over if used incorrectly.

✓ Do not apply any lateral force to the tripod.

✓ Always use a separate belay line that does not go through a high directional on the tripod.

T

Rescue Tripod and Winch

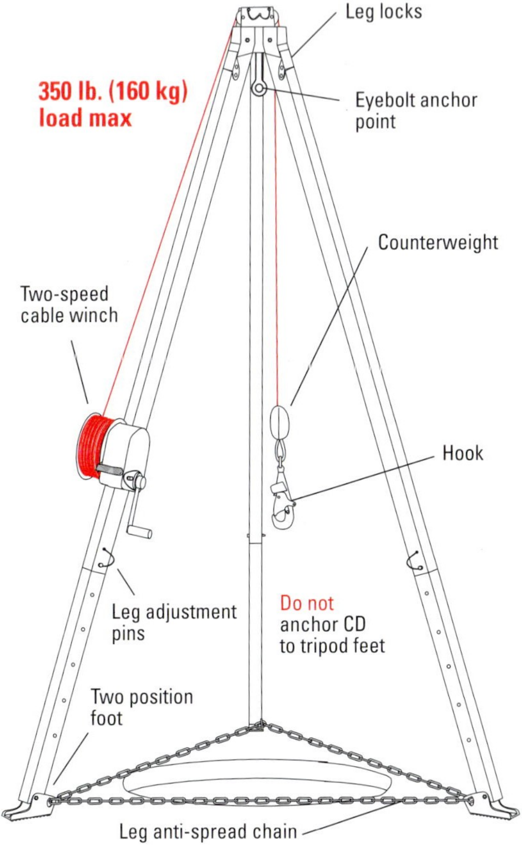

Leg locks

350 lb. (160 kg) load max

Eyebolt anchor point

Counterweight

Two-speed cable winch

Hook

Leg adjustment pins

Do not anchor CD to tripod feet

Two position foot

Leg anti-spread chain

Winch Cable Setup

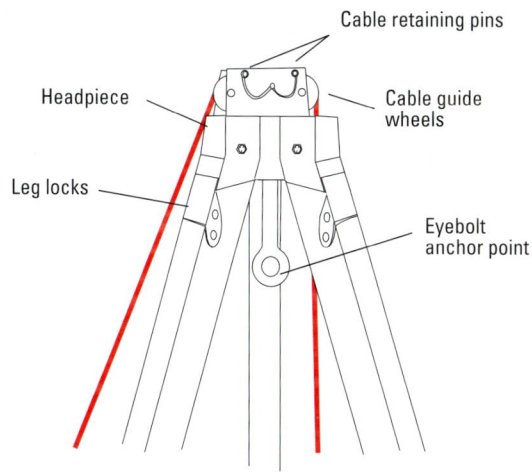

Cable retaining pins

Headpiece

Cable guide wheels

Leg locks

Eyebolt anchor point

Side View of Winch

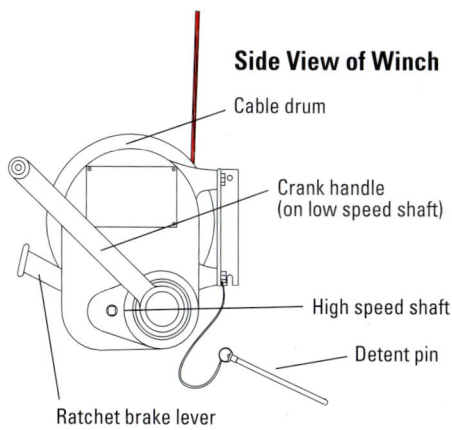

Cable drum

Crank handle (on low speed shaft)

High speed shaft

Detent pin

Ratchet brake lever

Rescue Tripod and Pulley System

Double sheave pulley

Pulley body

Cam lockout pin

Cam

Cord guide

Cam release cord

Tripod head

Anchor point

Confined space rescue pulley system (4:1 CD)

Cam release cord for lowering

Leg adjustment pins

Two position foot

Do not anchor CD to tripod feet

Leg anti-spread chain

Belay   **B**

Do not run belay through top of tripod

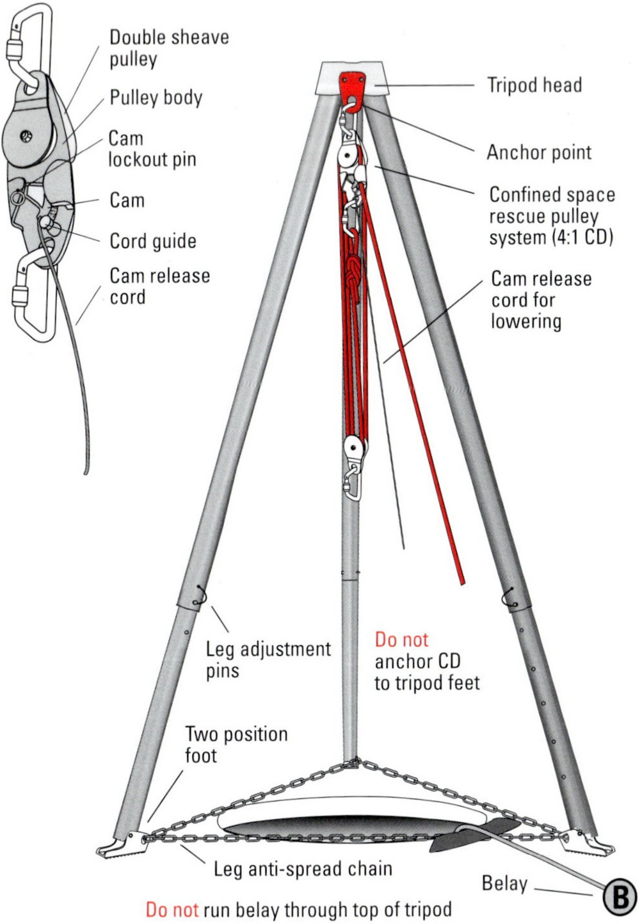

✓ Do not apply any lateral force to the tripod.

✓ Always use a separate belay line that does not go through a high directional on the tripod.

Aerial Apparatus as High Anchor Point

An aerial apparatus can create a safe and effective anchor point for a rescue system but can fail catastrophically if not done properly.

1. Spot apparatus as close to work area as possible.
2. Position tip of aerial directly over intended work area.
3. Check tip capacity chart on turn table to ensure that the aerial can support at least 500 lbs. (227kg) at that angle and extension.
4. If within capacity, return aerial to ground and rig anchor point and system.
5. Always keep haul force and system components in line with the center of the aerial. Lateral force on the aerial can cause structural failure.
6. Re-position aerial over work area.
7. Rig belay at separate anchor point.
8. Lift only one person at a time and never rotate, extend or retract the aerial with a person on the system. The aerial is to be used as an anchor point only!

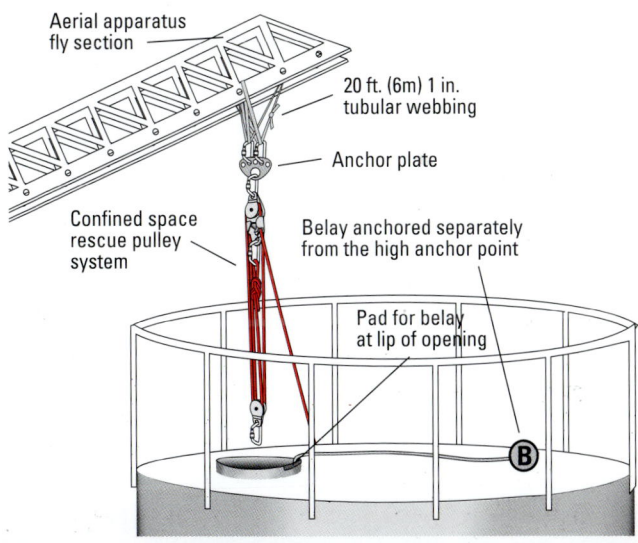

Aerial apparatus fly section

20 ft. (6m) 1 in. tubular webbing

Anchor plate

Confined space rescue pulley system

Belay anchored separately from the high anchor point

Pad for belay at lip of opening

Ⓑ

Aerial Apparatus as High Anchor Point

1. Create the anchor point on a standard aerial to distribute weight between both beams and at least 2 rungs.
2. Lay a 20 ft. (6m) webbing over both rails and pull up ends through adjacent rungs.
3. Tie with overhand bend.
4. Clip steel carabiner through both loops at point marked 1 and pull down.
5. Clip second steel carabiner through each loop marked 2 and pull down.

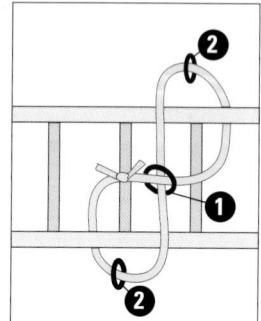

6. Attach anchor plate to carabiners as shown.

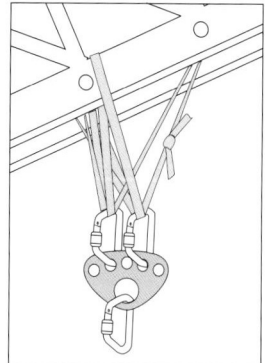

Swiftwater Rescue Command Checklist

Phase I: Size up
- [ ] Primary Assessment
  - [ ] Secure witness
  - [ ] Determine location, number & condition of victims
  - [ ] Identify immediate hazards
    - [ ] Water level rising or falling (mark waterline)
    - [ ] Surface loads (debris), hydraulics, hypothermia
- [ ] Secondary Assessment
  - [ ] Assess need for additional personnel and equipment
  - [ ] Assess need for additional equipment (boat)
- [ ] Rescue mode or recovery mode

Phase II: Pre-rescue operations
- [ ] Make general area safe (i.e., traffic and crowd control)
- [ ] Make rescue area safe
  - [ ] Assign safety officer
  - [ ] Assure team response to opposite bank
  - [ ] Personal protective equipment within 10 ft. (3m) of water
  - [ ] Assign downstream bag throwers
  - [ ] Assign upstream spotters
- [ ] Form incident action plan
  - [ ] Reach, throw, wade, row, go, helicopter
- [ ] Backup plans (i.e., paddle team with boat)
- [ ] Subject PFD and helmet
- [ ] Pre-rescue briefing

Phase III: Rescue operations
- [ ] Implement primary action plan
  - [ ] Make contact with subject
  - [ ] Apply protective equipment
  - [ ] Remove subject to safe area
- [ ] Transfer to ALS, consider hypothermia (p.153)

Phase IV: Termination
- [ ] Personnel Accountability Report (PAR)
- [ ] Collect water samples to assess contamination
- [ ] Consider decontaminating rescuers

A

Personal Protective Equipment

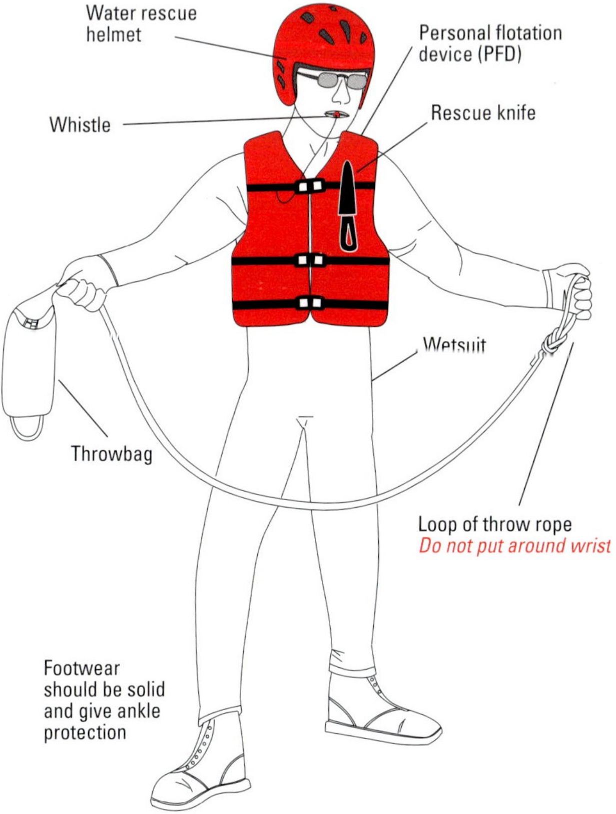

Water rescue helmet

Personal flotation device (PFD)

Whistle

Rescue knife

Wetsuit

Throwbag

Loop of throw rope
Do not put around wrist

Footwear should be solid and give ankle protection

✔ Water conducts heat away from the body 25 times faster than air of the same temperature.

Team Equipment

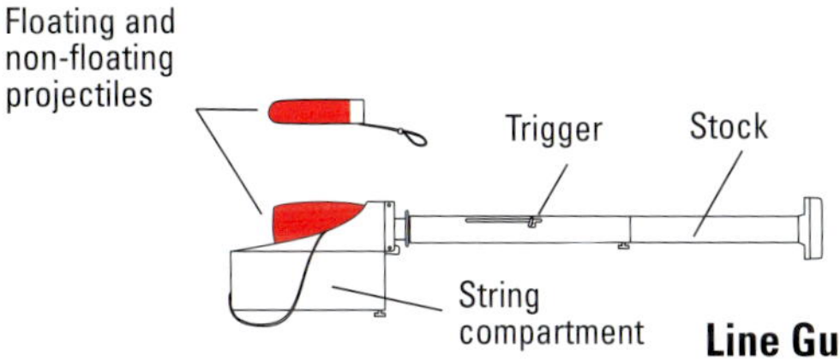

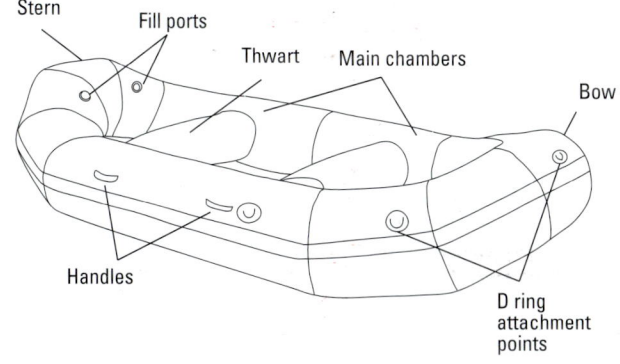

Floating and non-floating projectiles

Trigger

Stock

String compartment

Line Gun

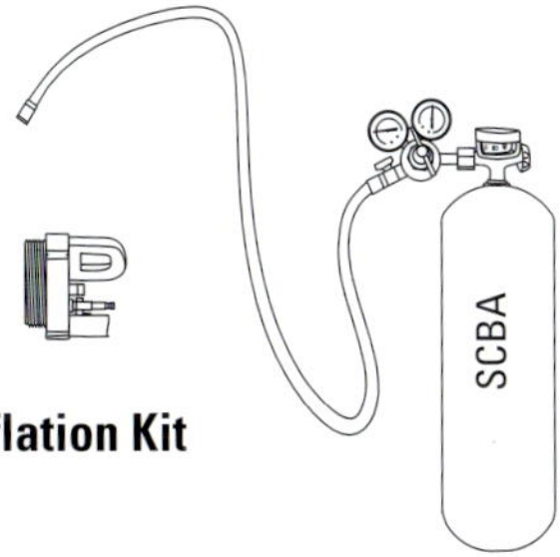

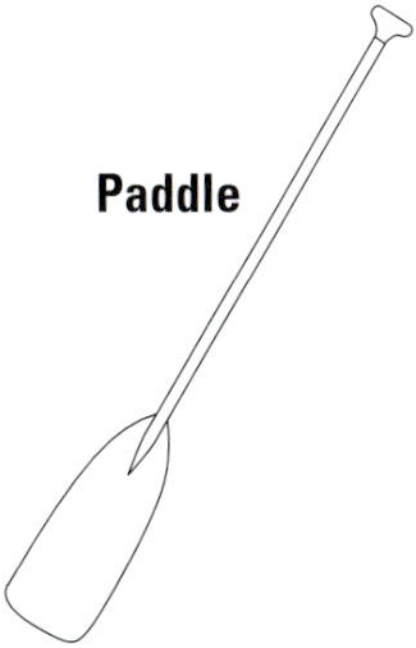

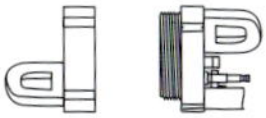

Hose Inflation Kit

SCBA

Paddle

Inflatable Rescue Boat

Stern

Fill ports

Thwart

Main chambers

Bow

Handles

D ring attachment points

Swiftwater Rescue Communications

Whistle blasts

One blast	= stop, look at me
Two blasts	= begin action agreed upon or indicated by whistle blower
Three blasts repetitive	= distress, need help

Hand signals

One arm in the air	= I need help
One hand on top of head	= I am ok

I need help!

I'm ok

Swiftwater Hazards

- Low head dam/hydraulics
- Strainers
- Hypothermia
- Floating debris
- Foot entrapment
- Stationary objects
- Panicked swimmers

River orientation
Rescuer facing down stream

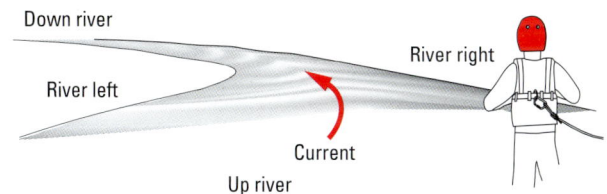

Methods of effecting a rescue in order of use/risk:

- Reach   **Low risk**
- Throw
- Wade
- Row
- Go
- Helo   **High risk**

✓ Do not go within 10 ft. (3m) of the water without a PFD on!

Safe Swimming Position

If you get swept away, assume safe swimming position and navigate with ferry angle.

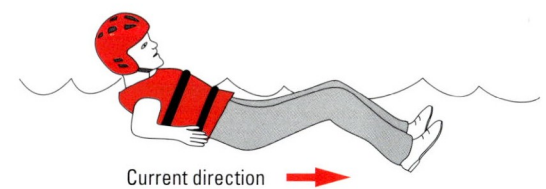

Current direction ➡

Ferry angle

1. Feet first, facing downstream.
2. Knees bent with feet slightly lower than butt.
3. Set proper ferry angle.
4. Angle body with head pointed 45 degrees toward the desired bank.
5. Stroke backward to help navigate.
6. Look for eddy and get set up well in advance.
7. Avoid strainers. If not possible to avoid, swim hard head first and attack up and over obstacle.

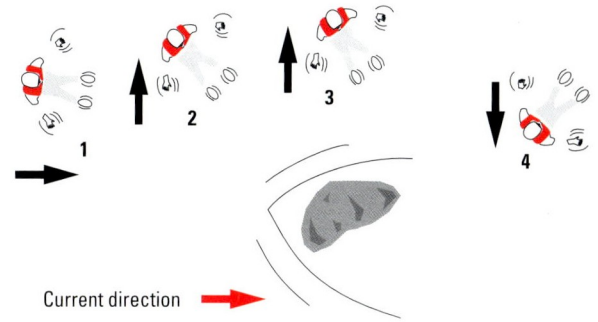

Current direction ➡

Shore-Based Rescue: Reach

Reach with an object such as a pikepole or paddle
1. Lay flat on the ground so as not to get pulled in.
2. Reach as far out as possible.
3. Yell to get the subject's attention.

Reach with an inflated fire hose in low head dam situations and bridge rescues
1. Connect as many sections of 2.5 in. diameter hose together as needed.
2. Cap one end.
3. Install the inflation manifold to the other end.
4. Tighten all couplings.
5. Assemble air bottle and regulator.
6. Inflate hose to 50 to 60 psi.
7. Form bight in end of hose and tie off.
8. Consider attaching PFD to end of hose.
9. Push hose out to victim.

Inflation manifold

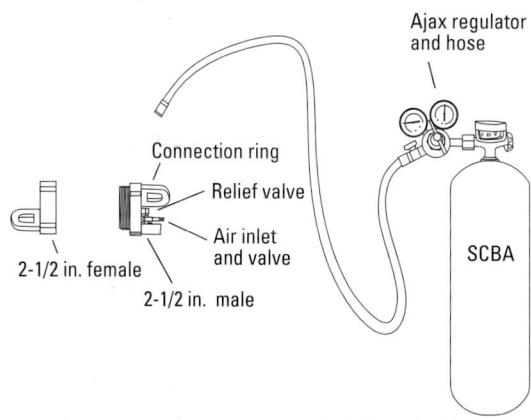

Ajax regulator and hose

Connection ring

Relief valve

Air inlet and valve

2-1/2 in. female

2-1/2 in. male

SCBA

Shore-Based Rescue: Reach

Reach subject with object

- Pike pole
- Paddle
- Tree branch
- Inflated fire hose

Other rescue options

- Flotation device tied to rope held by rescuers between river left and river right
- Boat on highline: track line must be up river from hydraulic and boat must be kept straight and away from face of hydraulic with downstream control lines
- Two boat access with downstream brake boat
- Remember aerated water reduces prop efficiency
- Lead boat must never contact face of hydraulic!

River wide hydraulic (low head dam)

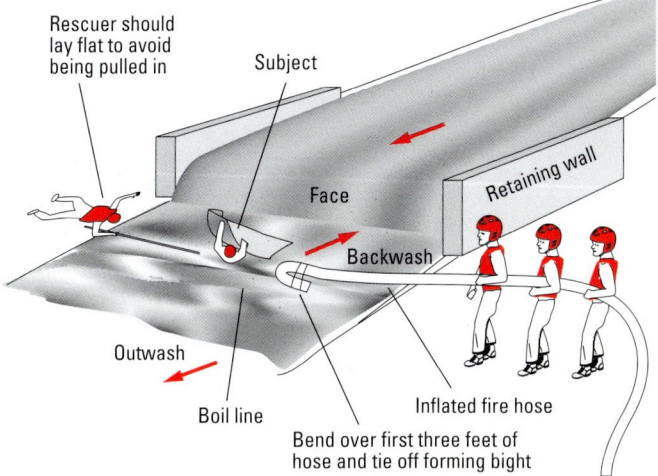

Rescuer should lay flat to avoid being pulled in

Subject

Face

Retaining wall

Backwash

Outwash

Boil line

Inflated fire hose

Bend over first three feet of hose and tie off forming bight

✓ Swimmers and boaters can identify this hazard from upriver by seeing a horizon line down river.

Shore-Based Rescue: Throw

1. Choose a strategic spot to set up to throw bag.
2. Get and keep eye contact with the subject .
3. Aim for the subject's head or slightly up river.
4. Make a strong underhand throw when the subject is in the target zone.
5. Carefully bring the subject to an eddy or the best landing spot you can find.
6. Be ready to make a second throw.

Remember

- Do not wade into current over your knees
- Consider a belay line for the rescuer throwing the bag if the shore is sloping and or if there is risk of the rescuer getting pulled in
- If the subject does not have a PFD on they will plane under in strong current. Try to give a moving belay and pull them in gradually

✓ Never count on the victim to participate in their own rescue.

Shore-Based Rescue: Throw

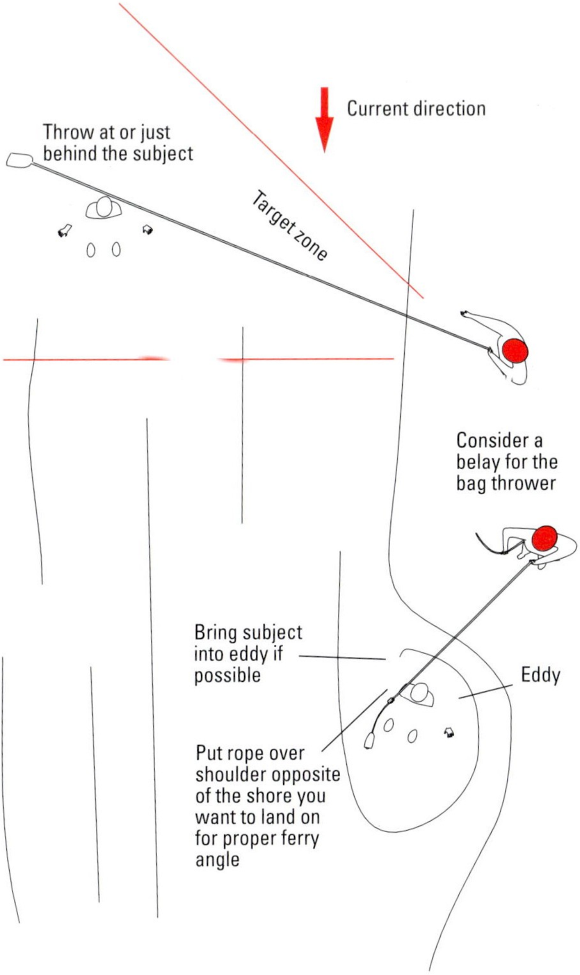

Current direction

Throw at or just behind the subject

Target zone

Consider a belay for the bag thrower

Bring subject into eddy if possible

Eddy

Put rope over shoulder opposite of the shore you want to land on for proper ferry angle

Shallow Water Crossing: Wade

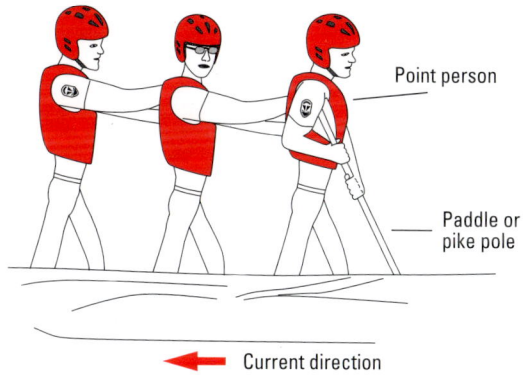

Point person

Paddle or pike pole

Current direction

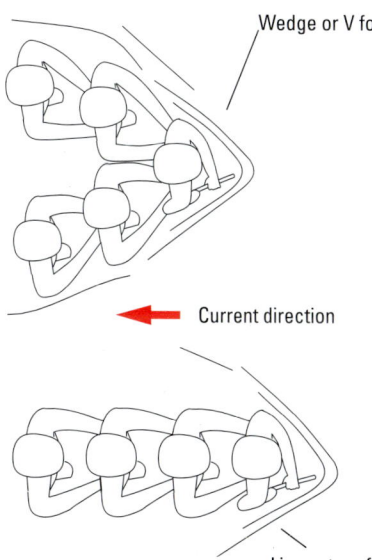

Wedge or V formation

Current direction

Line astern formation

- Do not enter current higher than knee deep
- Keep the formation headed straight into the current
- Support the person in front of you
- Get a solid foot placement each time you move your foot
- Do not rush
- Abort and return to shore before getting swept away

Shallow Water Crossing to a Vehicle

1. Do not enter current deeper than your knees.
2. Have upstream spotters to watch for floating debris.
3. Have downstream bag throwers as backup plan.
4. Secure vehicle with stabilization line if possible.
5. Do not follow the stabilization line, it leads to the reaction wave.
6. Take a PFD and helmet for each subject.
7. Keep the formation headed straight into the current (fig. A).
8. Abort the attempt if formation is not totally stable.
9. Move laterally to the rear of the vehicle, avoid the reaction wave.
10. Watch for instability of the vehicle.
11. If the vehicle appears stable, move up into the eddy (fig. B).
12. Get PFD and helmet correctly on each subject.
13. Assist one subject into the pocket of the V formation.
14. Move laterally to the safe bank (fig. C).
15. Repeat the process for additional subjects.

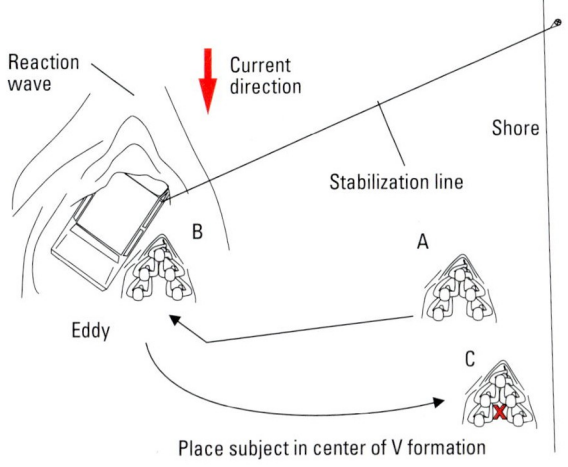

Reaction wave
Current direction
Shore
Stabilization line
B
A
Eddy
C

Place subject in center of V formation

Boat Operations: Row

Boat inflation procedure

1. Remove valve cap.
2. If valve stem is not flush with outside of valve, push in and turn 1/4 turn.
3. Insert fill nozzle and flow air.
4. Fill each main chamber to the point that it has shape.
5. Rotate around and gradually top off each chamber to insure equal pressure (floor and thwarts).
6. Final fill should give boat enough pressure to just dent tube with one knee.
7. Cap valves.

To deflate

1. Put one person on each valve, uncap and place finger on valve stem.
2. On mark, simultaneously push in valve stem on each main tube and turn 1/4 turn to keep open.
3. Open all other valves.
4. Fold boat, roll and stow in carry bag.

Halkey Roberts Valve

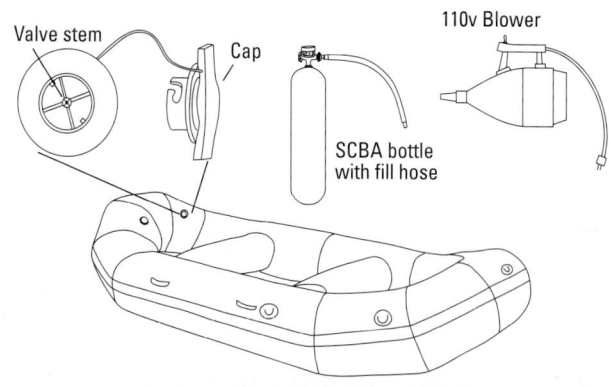

Valve stem

Cap

110v Blower

SCBA bottle with fill hose

Boat Operations: Row

Use a boat with paddle crew to:
- Paddle out to drifting subject
- Access hard to reach locations
- Have a backup plan to recover any rescuers swept away

Paddle crew procedures
1. Inflate boat.
2. Put one paddle for each rescuer plus one backup in boat.
3. Clip two throw bags into boat.
4. Assemble paddle crew.
5. Place boat in eddy or other suitable launch spot.
6. Paddle captain sits in back on raised stern.

Paddle captain is responsible for steering the boat.
Have two designated grabbers, others keep paddling.

Standard paddle commands

"Forward paddle" = All paddlers paddle forward.

"Back paddle" = All paddlers back paddle.

"Right turn" = Paddlers on right give one stroke back then continue forward. Left continues forward.

"Left turn" = Paddlers on left give one stroke back then continue forward. Right continues forward.

"Stop" = All paddlers stop paddling.

"High side" = Everyone move to the rising tube.

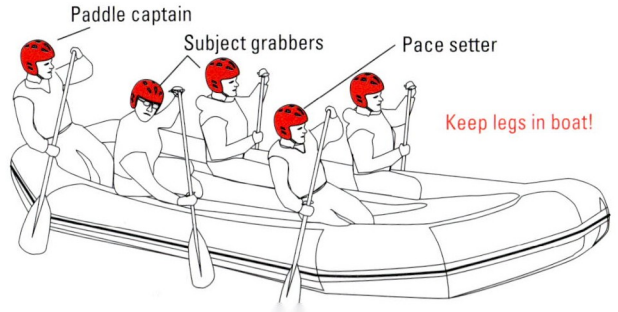

Paddle captain

Subject grabbers

Pace setter

Keep legs in boat!

Boat on Highline: Row

Boat on highline with movable control point is used to

- Precisely position boat in fast current
- Provide safe rescue platform
- Access low head dams
- Create a movable platform to catch drifting swimmers

Procedure

1. TSO assigns boat team, river right group and river left group.
2. Boat team inflates boat and rigs webbing bridle on front three D rings.
3. Remote side gets into position and locates suitable anchor.
4. Rescue side sends messenger line to remote side.
5. Remote side receives messenger line and pulls main rope across.
6. Anchor first line across and designate as track line.
7. Pull two additional lines across using track line.
8. Anchor second line and designate as remote control line.
9. Pull back using track line and anchor with ratchet prusik and PMP.
10. Pre-tension track line with 3:1 using no more than 1 puller.
11. Tie off track line with 3 ft. (1m) of slack between prusik and anchor.
12. Attach track pulley to track line and rig movable control point.
13. Attach rescue side control line to movable control point.
14. Boat can be rigged with no capability to lower, with 1:1 lower line, with 2:1 lower line or with pulley system controlled from within the boat.
15. 2:1 controlled from rescue side is recommended.
16. Post-tension track line with 3 pullers if needed.

Boat on Highline: Rigging

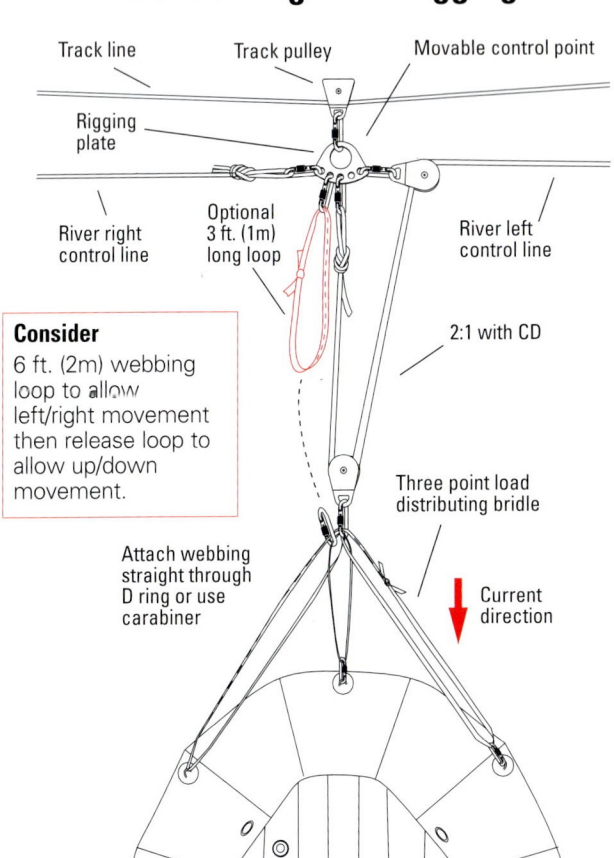

Track line

Track pulley

Movable control point

Rigging plate

River right control line

Optional 3 ft. (1m) long loop

River left control line

2:1 with CD

Consider
6 ft. (2m) webbing loop to allow left/right movement then release loop to allow up/down movement.

Attach webbing straight through D ring or use carabiner

Three point load distributing bridle

Current direction

✓ Drag on boat and tension on control lines will be severe in current faster than 10 ft./sec. (3m/sec.).

✓ Be prepared to put ratchet prusiks with attendants on control lines.

Boat on Highline: Operation

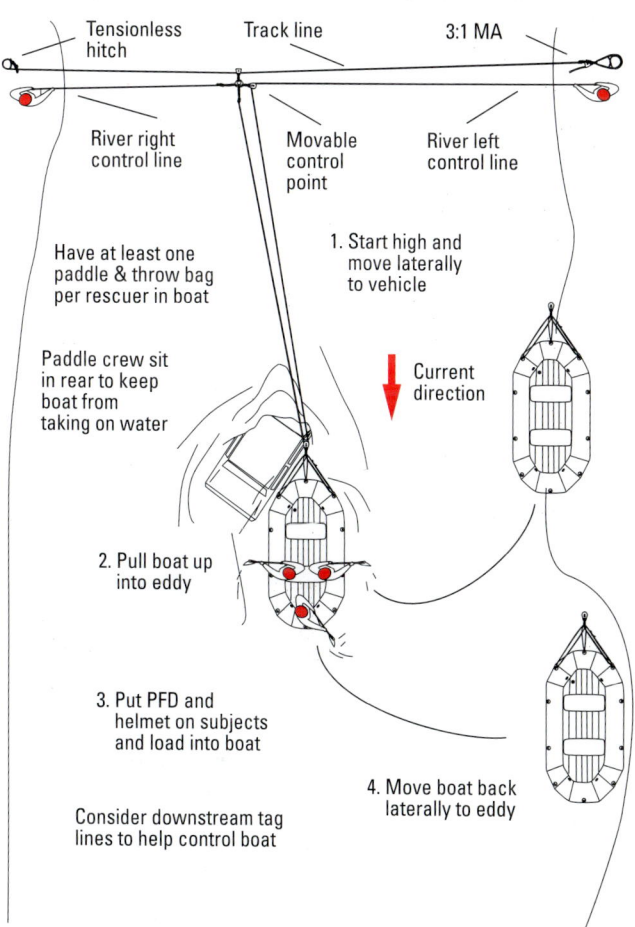

Tensionless hitch

Track line

3:1 MA

River right control line

Movable control point

River left control line

Have at least one paddle & throw bag per rescuer in boat

Paddle crew sit in rear to keep boat from taking on water

1. Start high and move laterally to vehicle

Current direction

2. Pull boat up into eddy

3. Put PFD and helmet on subjects and load into boat

4. Move boat back laterally to eddy

Consider downstream tag lines to help control boat

✓ The paddle captain must be prepared to cut the boat loose if the rope system puts boat in danger or if there is a need to chase a swimmer.

Boat on Highline:
Crew Signals to Shore Control

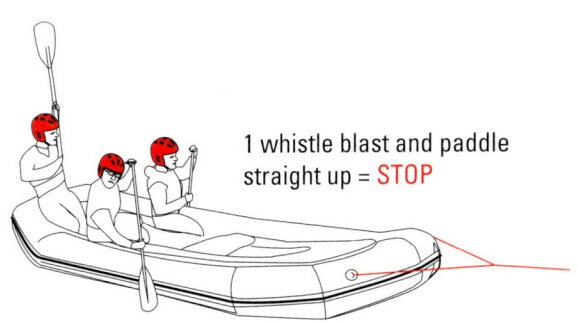

1 whistle blast and paddle
straight up = STOP

2 whistle blasts = move boat in direction
indicated by captain's raised paddle

River right

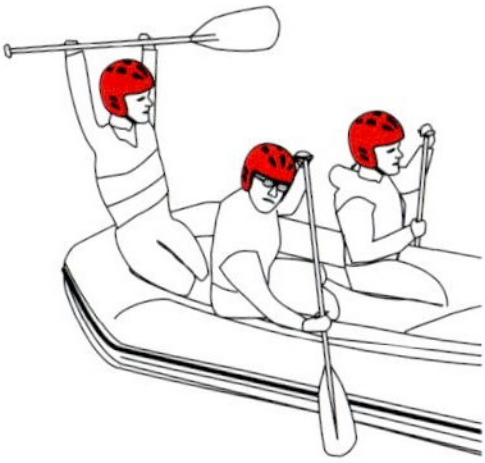

River left

Down river

Up river

Strong Swimmer Rescue: Go

Tethered rescuer special use rescue vest

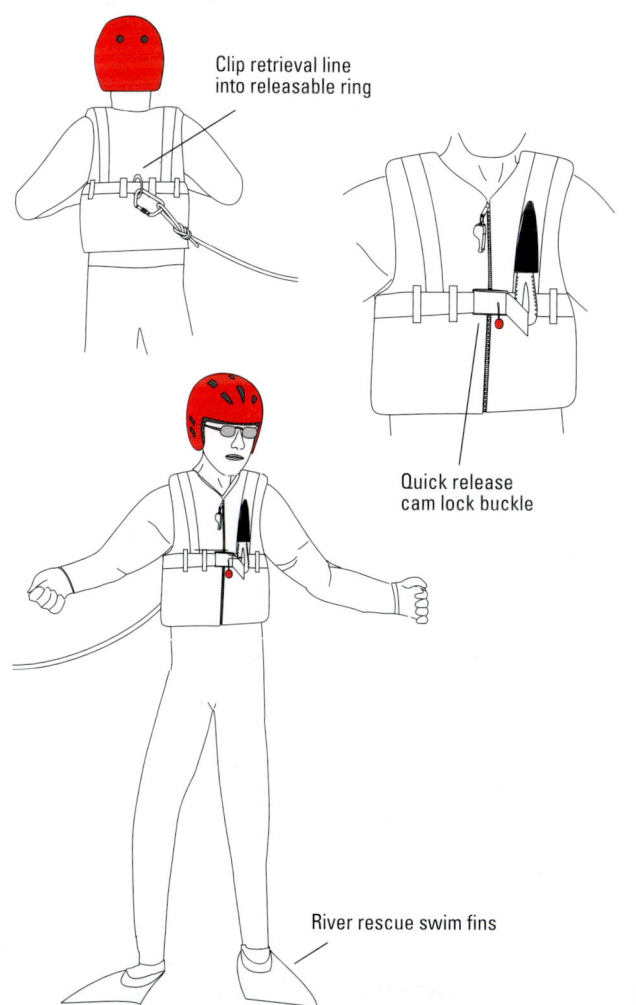

Clip retrieval line into releasable ring

Quick release cam lock buckle

River rescue swim fins

Tethered Strong Swimmer: Go

Indications for use
- As a backup plan
- To rescue a drifting subject who is out of throwbag range or unable to catch and use throwbag
- Useful range of about 150 ft. (45m)

Minimum requirements
- Recovery area clear of obstructions
- 1 Special purpose rescue vest
- 1 Pair river rescue fins
- 1 Strong swimmer with appropriate protective equipment
- 1 200 ft. (60m) rescue rope
- 4 Technical rescue technicians for support
- 2 Throw bags
- A good backup plan

Procedure
1. Set up at a downstream location with best advantage.
2. Strong swimmer dons river rescue fins and special purpose rescue vest.
3. 200 ft. (60m) rescue rope is connected by carabiner to releasable ring on back of rescue vest.
4. Stack approx. 20 ft. (6m) of rope down river from rescuer.
5. Position belay crew down river from rescue entry point.
6. Wait until subject is even with rescuer (fig. A).
7. Rescuer maintains eye contact with subject and performs shallow water dive entry (fig. B).
8. Rescuer swims aggressively to subject.
9. Belayers feed rope to prevent drag on rescuer and stand by for ready signal.
10. Rescuer holds subject with appropriate technique and indicates ready.
11. Belay team moves rescuer and subject to shore (fig. C).
12. If belayers are unable to bring rescuer in, rescuer can release from tether line and initiate backup plan (last resort).

✓ Never tie a rope around a rescuer. Only attach rope to a quick release system.

Tethered Strong Swimmer: Go

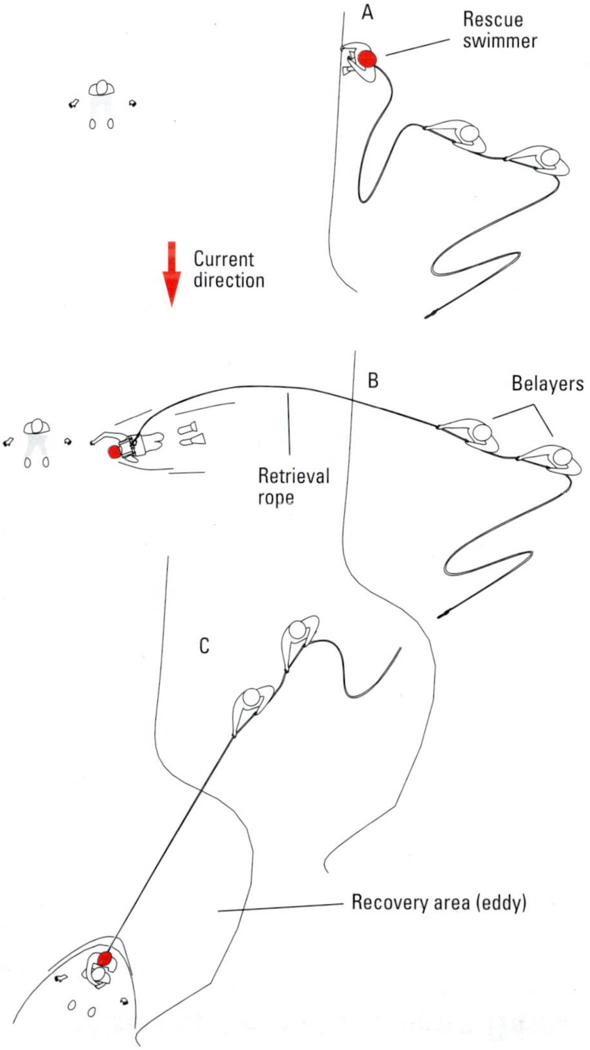

A

Rescue
swimmer

Current
direction

B

Belayers

Retrieval
rope

C

Recovery area (eddy)

Helicopters and Swiftwater: Helo

Helicopters are considered to have a high risk factor for swiftwater rescue operations. Prior to the use of a helicopter, all other options should have been ruled out due to the higher risk to rescuers or because they are incompatible with the situation. Rescuers should be the highest trained, strongest and best equipped available. A safety briefing must be conducted prior to starting operations. The pilot, as always, has the final word on go or no go.

Support role, less risk
- Transport rescuers across river
- Transport equipment across river
- Provide reconnaissance

Rescue role, highest risk
- Access vehicles and midstream objects
- Extract subjects via one-skid
- Insert rescuers onto objects via one-skid
- Insert rescue swimmers into water near subject
- Extract rescuers and subjects via short haul

! Follow local protocol for these high risk procedures!

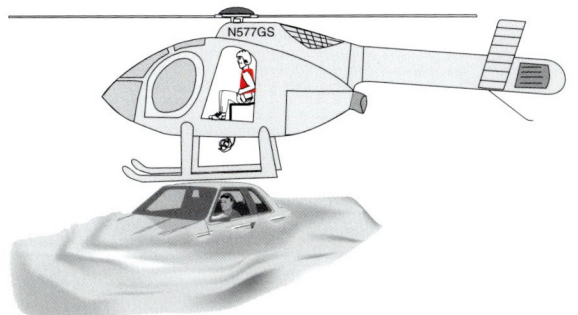

Trench Rescue Command Checklist

Phase I: Size-up

- [ ] Primary assessment
  - [ ] Secure witness or competent person
  - [ ] Identify immediate hazards
  - [ ] Location, number, condition of victims
  - [ ] Rescue or recovery
- [ ] Secondary assessment
  - [ ] Trench collapse [ ] yes [ ] no
  - [ ] Proper equipment and personnel on scene
  - [ ] Additional resources necessary: ventilation, shoring, retrieval system

Phase II: Pre-Entry Operations

- [ ] Make general area safe
  - [ ] Traffic control
  - [ ] Crowd control
  - [ ] Heavy equipment shut down
  - [ ] Establish zones: hot, warm, cold
- [ ] Make rescue area safe
  - [ ] Establish lobby control accountability
  - [ ] Secure hazards: gas, electric, utilities
  - [ ] De-water trench
  - [ ] Monitor atmosphere
  - [ ] Ventilate

Phase III: Rescue Operations

- [ ] Make trench lip safe
  - [ ] Assess spoil pile
  - [ ] Approach from ends
  - [ ] Place ground pads
- [ ] Make trench safe
  - [ ] Access/egress ladders less than 50 ft. (15m) apart
  - [ ] Protective system; sloping, hydraulic, timber, other
  - [ ] Create safety zones
  - [ ] Remove dirt: extend safety zones.

A

☐ Victim assessment.
    ☐ Treatment in trench, see crush syndrome (p. 151)
    ☐ Patient packaging
    ☐ Retrieval system/extrication
    ☐ Transfer to treatment sector

Phase IV: Termination
☐ Personnel Accountability Report (PAR)
☐ Remove tools and equipment
☐ Remove protective system
    ☐ Last in - first out
☐ Secure scene
☐ Consider debriefing
☐ Call OSHA

A

Soil Types

Type A: cohesive soils with an unconfined, compressive strength of 1.5 ton/sq. ft. (tsf) (144 kPa) or greater (most stable)

Type B: cohesive soil with an unconfined compressive strength greater than 0.5 tsf (48 kPa) but less than 1.5 tsf (144 kPa)

Type C: cohesive soil with an unconfined compressive strength of 0.5 tsf (48 kPa) or less (least stable)

Unconfined compressive strength: the load per unit area at which a soil will fail in compression.

In a rescue situation, soil types are considered to be worst case scenario (type C) and shoring should be constructed accordingly.

Trench Incident Site Setup

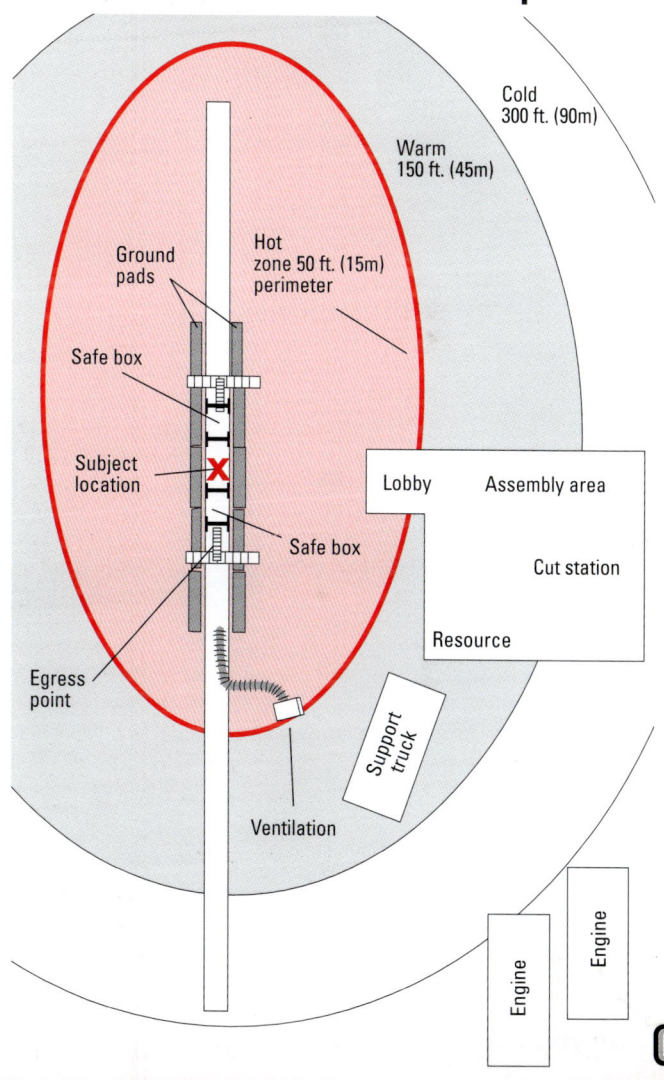

Cold
300 ft. (90m)

Warm
150 ft. (45m)

Hot
zone 50 ft. (15m)
perimeter

Ground
pads

Safe box

Subject
location

Safe box

Egress
point

Ventilation

Lobby   Assembly area

Cut station

Resource

Support
truck

Engine

Engine

Trench Definitions and OSHA Regulations 🅰

- Any trench 4 ft. (1.2m) deep or greater must have a means of egress within 25 ft. (7.5m) of any worker
- Any trench with a hazardous atmosphere or a potential hazardous atmosphere that is 4 ft. (1.2m) deep or greater must be monitored prior to employee entry
- An occupied trench 5 ft. (1.5m) deep or greater must have an approved protective system to protect employees from cave-ins
- Protective systems shall be placed from the top working down and removed from the bottom working up so as to protect the employee during construction or removal
- During rescue operations all soil should be considered type C and protective systems and practices shall be used accordingly

- *The timber shoring system illustrated in this guide is designed by a registered professional engineer for the Phoenix Fire Department, and any agency wishing to use this system or a similar system must establish an agreement with a registered professional engineer*

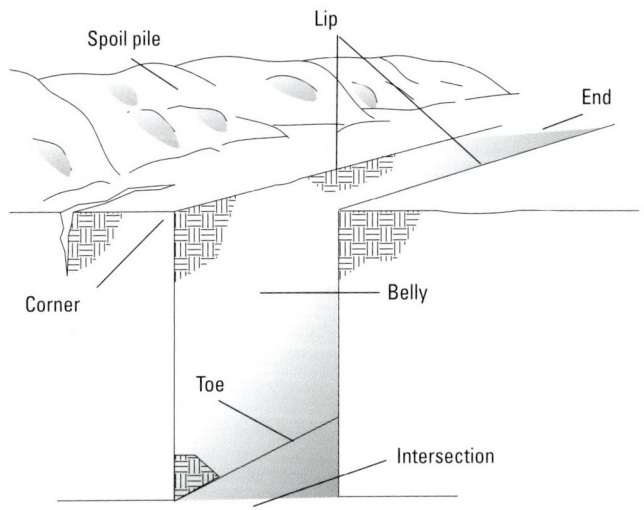

Trench Hazards and Causes of Collapse

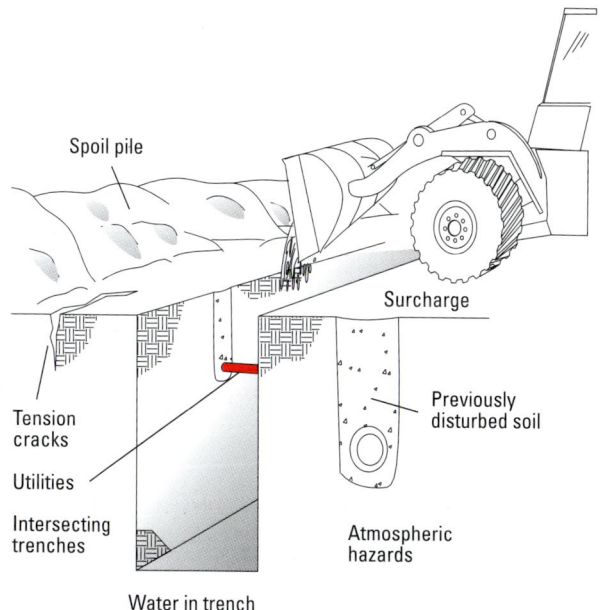

Spoil pile

Surcharge

Tension cracks

Utilities

Intersecting trenches

Previously disturbed soil

Atmospheric hazards

Water in trench

Remember:

- Soil weighs approximately 100 lbs./ft³ (50kg/.5m³) per cubic ft. and 3000 lbs./yd³ (1400kg/m³)
- If an initial collapse occurred, secondary collapse is highly likely
- Consider the possibility of a hazardous atmosphere in a trench
- Exposed utilities should be supported in place

✓ Do not enter an unprotected trench for any reason!

Hydraulic Speed Shore® System

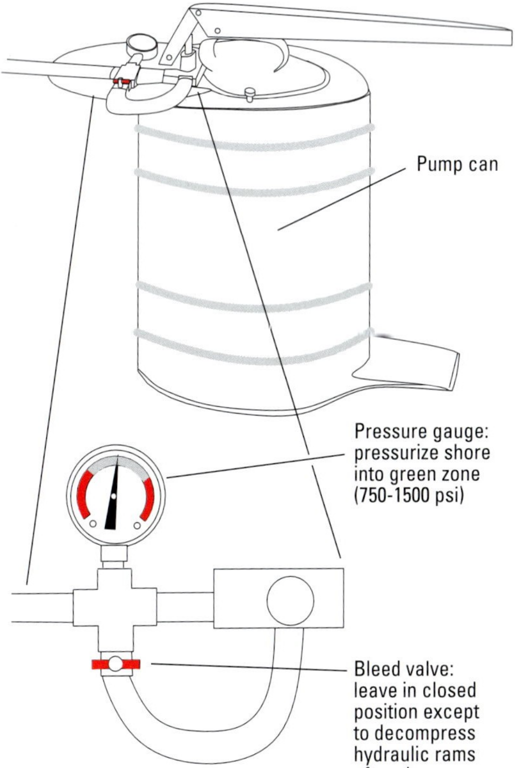

Pump can

Pressure gauge:
pressurize shore
into green zone
(750-1500 psi)

Bleed valve:
leave in closed
position except
to decompress
hydraulic rams
after shore
removal

✔ Pressurize all shores to a consistent pressure within the green zone.

Speed Shore Installation

1. Measure trench width and depth and then select suitable shore.
2. Lower two sheets of form sheeting into trench at designated location and hold.
3. Connect pump can pressure hose to inlet port on upper cylinder assembly.
4. Place pump can bleed valve in closed position.
5. Attach T handle to handle on lower vertical rail.
6. Grasp handle of other vertical rail and lower shore into trench.
7. Release upper vertical rail and let it drop into position.
8. Hold shore in position and pump up to operation pressure (green).
9. Do not open bleed valve.
10. Remove hose from coupling with end of T handle.

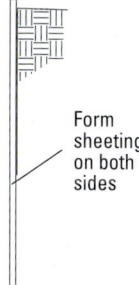

For irregular or sloping walls a single cylinder rescue shore can be used

Ground pad

Coupling

Form sheeting on both sides

✔ Avoid the use of hydraulic shores near trench intersection corners.

Aluminum Hydraulic Speed Shore

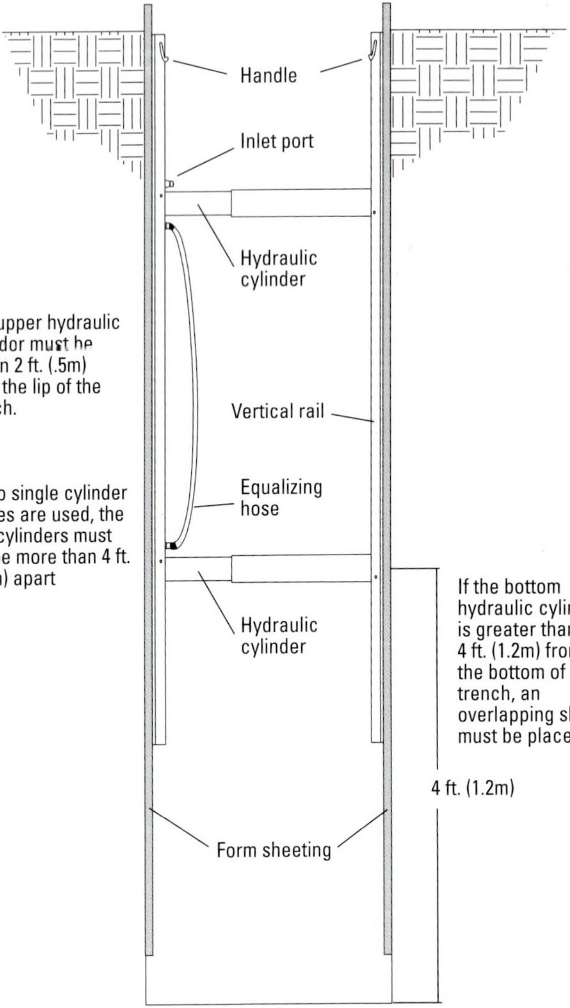

Handle

Inlet port

Hydraulic
cylinder

The upper hydraulic
cylinder must be
within 2 ft. (.5m)
from the lip of the
trench.

Vertical rail

Equalizing
hose

If two single cylinder
shores are used, the
two cylinders must
not be more than 4 ft.
(1.2m) apart

Hydraulic
cylinder

If the bottom
hydraulic cylinder
is greater than
4 ft. (1.2m) from
the bottom of the
trench, an
overlapping shore
must be placed.

4 ft. (1.2m)

Form sheeting

Pneumatic Shore Placement

Pneumatic shores can be properly installed from outside the trench initially. Try to protect as much of the trench as possible using this technique before entering.

1. Clean lip of trench and place ground pads around work area.
2. Measure trench depth and width.
3. Select suitable size shores and lay out air control system.
4. Assemble 3 sets of timber panels.
5. Install first set of panels as close to subject location as possible.
6. Attach utility rope and air hose to first shore (set up additional shores with rope and air hoses). (continued)

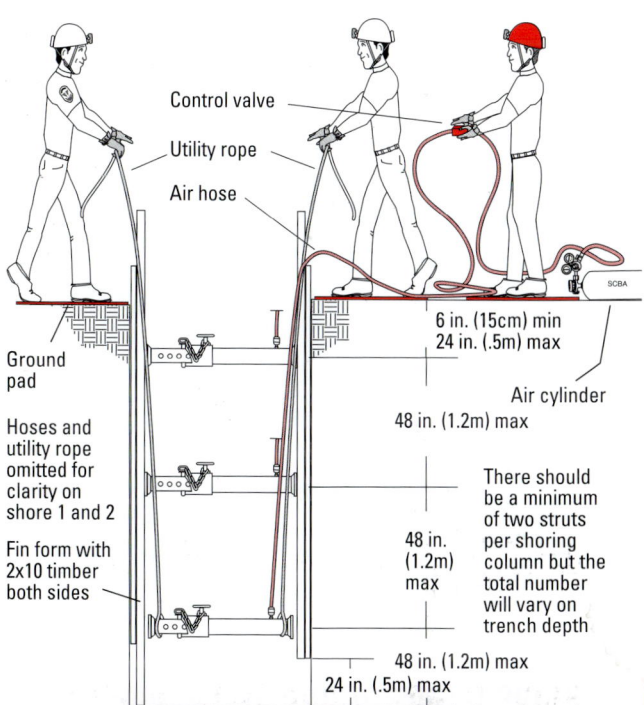

Control valve

Utility rope

Air hose

Ground pad

Hoses and utility rope omitted for clarity on shore 1 and 2

Fin form with 2x10 timber both sides

6 in. (15cm) min
24 in. (.5m) max

Air cylinder

48 in. (1.2m) max

There should be a minimum of two struts per shoring column but the total number will vary on trench depth

48 in. (1.2m) max

48 in. (1.2m) max
24 in. (.5m) max

SCBA

Pneumatic Shore Placement

7. Lower first shore into trench and position between 6 in. (15cm) and 24 in. (61cm) of the lip.

8. Ensure shore is level, then pressurize appropriately.

9. Tie off utility rope to top of panel and place second shore within 48 in. (122cm) of first shore.

10. Place third shore and pressurize.

11. Install second set of panels within 4 ft. (1.2m) on center of first set and repeat shore process.

12. Install third set of panels within 4 ft. (1.2m) on center on the other side of the first set of shores.

13. Set pins on shores and tighten all collars.

14. Nail or screw each shore base plate to timber.

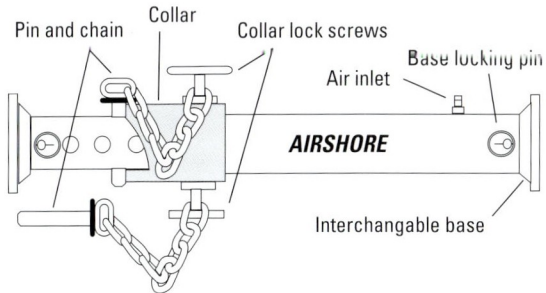

Pneumatic shore key points are

- Install shores from top down or according to manufacturer
- Pressurize to manufacturers specifications per soil conditions
- Tighten collars and screw or nail base to timber
- Always refer to tabulated data provided by the manufacturer for correct spacing

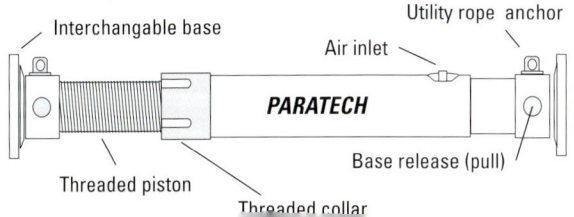

Timber Shore Step-by-Step

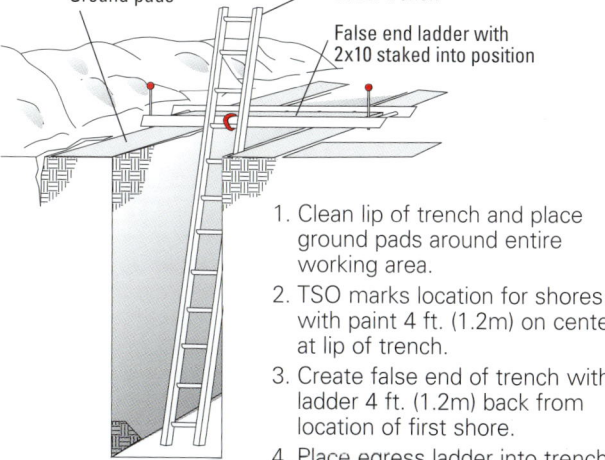

Ground pads

Egress ladder 36 in. (1m) out of trench

False end ladder with 2x10 staked into position

1. Clean lip of trench and place ground pads around entire working area.
2. TSO marks location for shores with paint 4 ft. (1.2m) on center at lip of trench.
3. Create false end of trench with ladder 4 ft. (1.2m) back from location of first shore.
4. Place egress ladder into trench and secure to end ladder.
5. Assemble vertical uprights by placing one piece of form sheeting on top of 2x10 (may be pre-assembled).
6. Position with 1 ft. (.3m) of 2x10 exposed on each end and nail into place.
7. Flip over and nail bottom joist hanger no greater than 2 ft. (.5m) from end.
8. Measure trench and determine best location for last joist hanger so that it will be inside trench.
9. Joist hangers must not be greater than 3 ft. (1m) apart for a trench less than 8 ft. (2.5m) deep.
10. Assemble opposite vertical upright in the same manner but do not put on joist hangers. (continued)

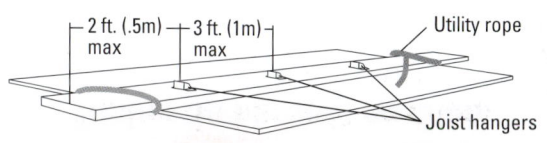

2 ft. (.5m) max — 3 ft. (1m) max — Utility rope

Joist hangers

Timber Shore Step-by-Step

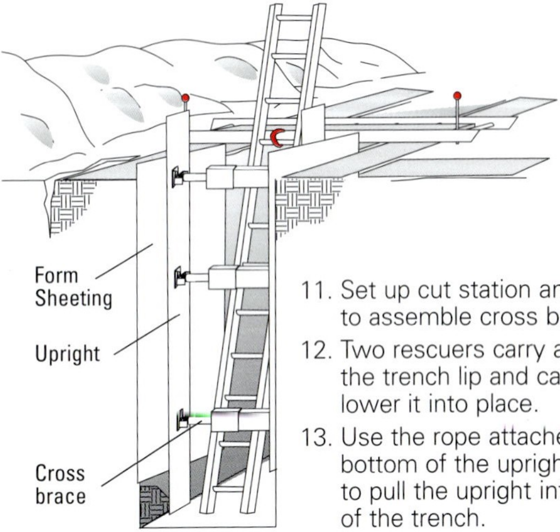

Form Sheeting

Upright

Cross brace

11. Set up cut station and prepare to assemble cross braces.

12. Two rescuers carry an upright to the trench lip and carefully lower it into place.

13. Use the rope attached to the bottom of the upright assembly to pull the upright into the toe of the trench.

14. When both uprights are in position, the builder climbs down the ladder no more than waist deep and measures the distance from upright to upright.

15. The builder calls out the distance in inches and the cut station subtracts 10 in. (25cm) and make the cut.

16. The assembly crew adds the 4x4 to a screw jack and secures it with one nail.

17. A utility rope is clipped to the assembled cross brace and it is handed to the builder.

18. The builder places the 4x4 in the top joist hanger and begins to expand the screw jack.

19. Place a nail in the base of the screw jack to secure it to the upright.

20. Repeat the process from the top down, until all cross braces are in position.

21. A final tightening is done on each cross brace.

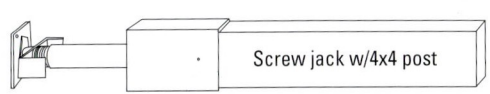

Screw jack w/4x4 post

Timber Shore Step-by-Step

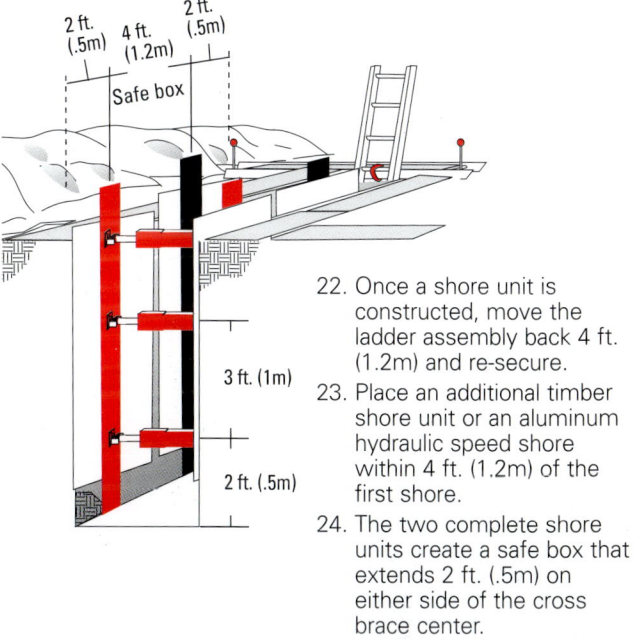

2 ft. (.5m) 4 ft. (1.2m) 2 ft. (.5m)

Safe box

3 ft. (1m)

2 ft. (.5m)

22. Once a shore unit is constructed, move the ladder assembly back 4 ft. (1.2m) and re-secure.

23. Place an additional timber shore unit or an aluminum hydraulic speed shore within 4 ft. (1.2m) of the first shore.

24. The two complete shore units create a safe box that extends 2 ft. (.5m) on either side of the cross brace center.

25. An additional shore unit would extend the safe zone provided it is placed within 4 ft. (1.2m) on center.

26. Once a safe box is constructed, dirt removal and patient stabilization can occur.

27. Move as much dirt laterally as is possible then move dirt vertically with buckets, if required.

28. Use several buckets and only fill half full.

29. Be prepared for crush syndrome (p.151).

30. Focus on patient removal as first priority and ALS treatment as second priority.

✓ You must be prepared to modify timber shore to fit the irregular shape and depth of a collapse area.

Wale System

A wale system is used for protecting intersecting trenches and other difficult to shore locations.

1. Follow all steps required to make the trench safe and to construct protected safe zones for workers to enter the trench.
2. Construct a safe box on all sides of the intersection or area to be protected.
3. Measure distance on center between safe boxes (must not exceed 6 ft. (2m) per engineer specifications).
4. Nail joist hangers on 6x6 wales at specified distance apart.
5. Attach utility rope to each end of wale and lower into trench, feeding between protective system components.
6. Place top wale within 18 in. (.4m) of trench lip (plan ahead).
7. Place second wale within 3 ft. (1m) on center of first wale.
8. Measure distance between wale and opposite vertical upright.
9. Place cross braces as noted in earlier procedure.
10. Slide sheeting between wale and unprotected trench area.
11. This system is for trenches between 5 ft. and 8 ft. (1.5 to 2.5m) deep. Deeper trenches require larger walers 3 ft. (1m) on center up to 11 ft. (3.3m) deep and 2 ft. (.5m) on center between 11 ft. and 15 ft. (3.3 to 4.5m).
12. If trench is too narrow to use screw jacks, use wedges.

✓ Standing water in trench is inherently unsafe and must be removed prior to shoring.

Wale Setup

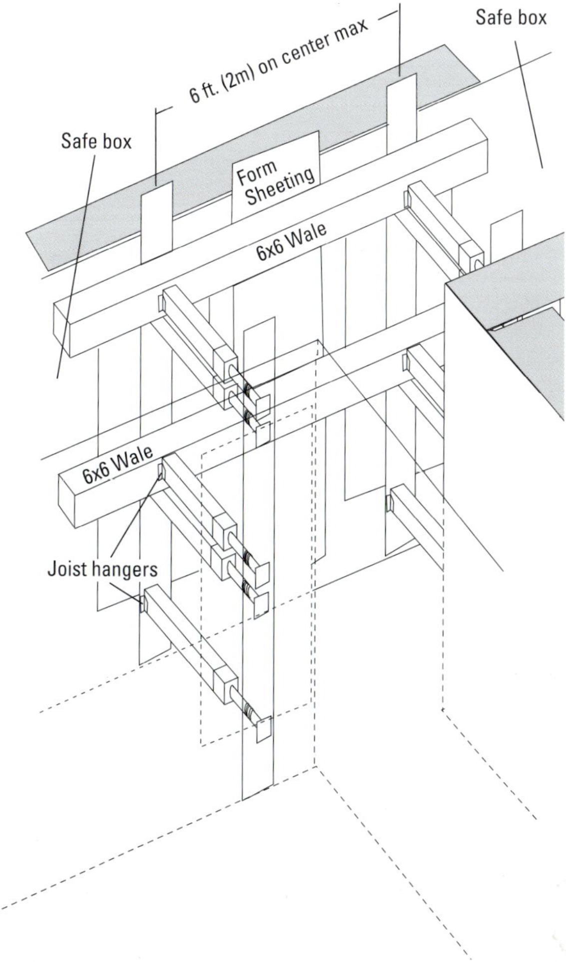

Safe box

Safe box

6 ft. (2m) on center max

Safe box

Form Sheeting

6x6 Wale

6x6 Wale

Joist hangers

Structural Collapse Command Checklist Ⓐ

Phase I: Size up
☐ Primary assessment
  ☐ Secure witness or reporting party (RP)
  ☐ Determine location, number & condition of victims
  ☐ Determine location and number of buildings involved
  ☐ Rescue mode or recovery mode
☐ Secondary assessment
  ☐ Type of occupancy (business, mercantile, assembly)
  ☐ Building construction type
  ☐ Assess hazards (secondary collapse, gas, electric)
  ☐ Assess need for additional personnel (search dogs, Red Cross, structural engineer, surgeon, FEMA)
  ☐ Assess need for additional equipment (100 ton crane, heavy equipment, lumber in quantity)

Phase II: Pre-rescue operations
☐ Make general area safe (i.e., traffic & crowd control)
  ☐ Establish transportation corridor
☐ Make rescue area safe (secure utilities)
  ☐ Establish perimeter (lobby control)
  ☐ Assign safety officer
  ☐ Personal protective equipment
  ☐ Establish victim staging area (accountability)
  ☐ Establish treatment area
☐ Remove all non-essential personnel from rescue area
☐ Establish building triage team
☐ Establish action plan for building search team
☐ Establish action plan for rescue team
☐ Rescue briefing

Phase III: Rescue operations
☐ Remove surface victims
☐ Implement search and rescue action plan
☐ Medical (treatment, transport)

Phase IV: Termination
☐ Personnel Accountability Report ☐ Remove equipment
☐ General debris removal ☐ CISD

Task Level Checklist

Assessment

- ☐ Reconnoiter entire site
    - ☐ Determine structure type
    - ☐ Interview neighbors, survivors to determine how many potential victims and points last seen
    - ☐ Obtain building plan or draw crude plan
        - ☐ Probable location of voids
        - ☐ Best access
        - ☐ Basements
- ☐ Search specialists re-assess building to re-identify hazards
- ☐ Prioritize site and make risk management profile
    - ☐ Structural engineer
        - ☐ Determines level of access to structure
        - ☐ Determines appropriate shoring
    - ☐ Plan shoring at accesses, and or use most efficient access
    - ☐ Determine condition of basement
    - ☐ Avoid falling hazards unless they can be removed or shored

Search

- ☐ Initial search
    - ☐ Conduct visual search
    - ☐ Conduct callout/listen search
    - ☐ Search from safe stable areas into unstable areas
    - ☐ Explore horizontal openings with great care
- ☐ Advanced search
    - ☐ Use search dogs with send out as far as possible
        - ☐ Check alerts with second dog
    - ☐ Use listening seismic finders, if available
    - ☐ Use search cam and thermal imager on high probability areas
    - ☐ Explore existing vertical shaft openings
    - ☐ Re-prioritize site vs. location of potential live victims

[A]

Access

- [ ] Rescue squad/rescue operations
  - [ ] Establish shoring/extrication plan
  - [ ] Cut team
    - [ ] 5 technicians optimum
    - [ ] Lumber supply
    - [ ] Build cut station
  - [ ] Rescue team
    - [ ] Gather all tools
    - [ ] Use appropriate protective equipment
    - [ ] Measure area to be shored and request material from cut team
  - [ ] Initial shoring for access
  - [ ] Build shores in safe area if possible
- [ ] Selected cutting and removal
  - [ ] Cut vertical openings and re-search, re-check with dogs
  - [ ] Avoid unshored overhead structures
  - [ ] Recheck all shoring after cutting and removal
  - [ ] Stabilize area at victim to give aid

Extricate

- [ ] Assemble medical specialist at patient location with treatment and extrication gear
  - [ ] Conduct patient assessment and triage
  - [ ] Provide appropriate level of treatment
  - [ ] Package patient for extrication
- [ ] Remove patient to treatment area

All clear on structure

- [ ] Re-assess survivability profile considering:
  - [ ] Elapsed time
  - [ ] Condition of remaining structure
- [ ] Declare all clear on entire structure or as much of structure as possible when appropriate

All shoring systems in this guide are for reference only and are subject to modification and approval by the on-site structural engineer.

FEMA Task Force
Search and Rescue Marking System

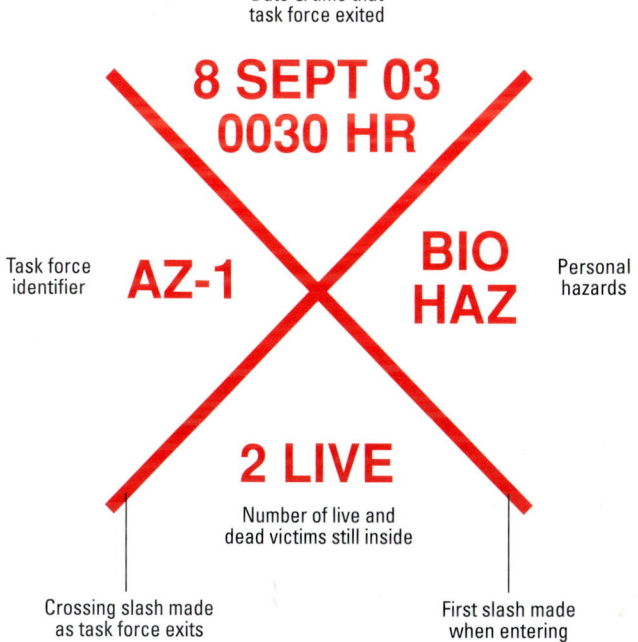

Date & time that
task force exited

8 SEPT 03
0030 HR

Task force
identifier

AZ-1

BIO
HAZ

Personal
hazards

2 LIVE

Number of live and
dead victims still inside

Crossing slash made
as task force exits

First slash made
when entering

FEMA Task Force Building Marking System

A

Structure and hazard evaluation

Structural specialist makes a 2′ x 2′ box on building adjacent to most accessible entry. This is done after doing hazards assessment and filling out hazards assessment form. Box (illustrated below) is spray painted with international orange paint and marked as follows:

Structure is relatively safe for SAR operations. Damage is such that there is little danger of further collapse (can be pancaked building).

Structure is significantly damaged. Some areas might be relatively safe, but other areas may need shoring, bracing or removal of hazards.

Structure is not safe for rescue operations and might be subject to sudden collapse. Remote search operations can proceed at significant risk. If rescue operations are undertaken, safe haven areas and rapid evacuation routes should be created.

Arrow located next to the marking box indicates the direction of safest entry to the structure.

Indicates hazmat condition in or adjacent to structure. SAR operations normally will not be allowed until condition is better defined or eliminated.

Example

8 Sept 03 0030 hrs
HM - natural gas
AZ-1

Until gas is turned off

Cut Station Setup and Operation

Cut team positions

1 Cutter
2 Feeders
1 Runner
1 Layout person

Tools

2- Carpenter kits
1- 10 1/4 in. circular saw
1- Chain saw with supplies
1- K12 saw with supplies
110 volt power supply, 20A preferred
200 ft. (60m) Extension cords

1. Cut one: 4 ft. x 4 ft. (1.2x1.2m) sheet 3/4 in. plywood
   Cut four: 4 ft. (1.2m) 2x4 lacing pieces
   Cut four: 2 ft. (.5m) 4x4 legs
2. Nail 2x4 lacing at top of first two 4x4 legs
3. Repeat with second set of legs
4. Set both sets of legs on end and join with 2x4 lacing
5. Flip legs over and add last piece of 2x4 lacing
6. Nail 4 ft. x 4 ft. (1.2x1.2m) sheet of plywood on burying nails
7. Nail 2x4 guides at indicated intervals

Cut station operation

9. The cut station needs material to function. Request from resource:
   - 4 sheets plywood
   - 12-36 4x4 or 6x6 posts (qty. and length depend on job)
   - 12-36 2x6 (qty. and length depend on job)
   - 12-36 2x4 (qty. and length depend on job)
10. Begin cutting gusset plates and wedges immediately.
    - Cut 64 1 ft. x 1 ft. (.3x.3m) gusset plates
    - Cut 24 sets 18 in. (.4m) 4x4 wedges
    - Cut 24 sets of 12 in. (.3m) 2x4 wedges
11. Cut wedge sets by placing stock in guide on cut table.
12. Layout cut marks on entire piece of stock with marks to match sets.
13. Feeder holds stock while cutter makes first diagonal cut.
14. Cut second half of wedge off and repeat process.
15. Runner takes material to shoring team and gets cut list.

Cut Station

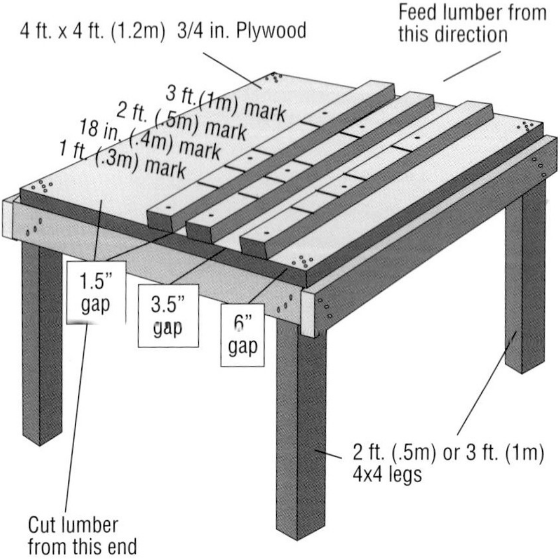

Feed lumber from this direction

4 ft. x 4 ft. (1.2m) 3/4 in. Plywood

3 ft. (1m) mark
2 ft. (.5m) mark
18 in. (.4m) mark
1 ft. (.3m) mark

1.5" gap

3.5" gap

6" gap

2 ft. (.5m) or 3 ft. (1m) 4x4 legs

Cut lumber from this end

18 in. (.4m) 4x4 Wedge layout cut sequence

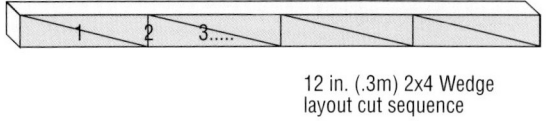

1   2   3.....

12 in. (.3m) 2x4 Wedge layout cut sequence

Material Capacities and Weights

Material Capacities

Post Shores	
Height of post	Capacity
4x4 8 ft. (2.5m)	8000 lbs.
4x4 10 ft. (3m)	5000 lbs.
4x4 12 ft. (3.6m)	3500 lbs.
6x6 8 ft. (2.5m)	20,000 lbs.
6x6 10 ft. (3m)	12,000 lbs.
6x6 12 ft. (3.6m)	7500 lbs.

Wood Cribbing	
Size	Capacity
2-4x4	24,000 lbs.
3-4x4	55,000 lbs.
2-6x6	60,000 lbs.
3-6x6	136,000 lbs.

- Limit height to triple width
- Overlap corners by 4 inches
- Bottom layer should be solid

Box Shores	
Height of post	Capacity
4-4x4 up to 16 ft. (4.8m)	30,000 lbs.
4-6x6 up to 24 ft. (7.3m)	72,000 lbs.

Miscellaneous Strengths	
Material	Capacity
4x4 screw jack	use 4x4 capacity
16d nail	150 lbs. shear

Unit Weights

Common Material	
Type	Weight lbs.
Concrete	150 pcf
Masonry	125 pcf
Wood	35 pcf
Steel	490 pcf
Masonry Rubble	10 psf/inch

Common Construction	
Type	Weight lbs.
Concrete floors	90-150 psf
Steel beam/ concrete deck	50-70 psf
Wood floors	10-25 psf

- Add 10-20 psf for wood or metal stud wall each level
- Add 10 psf for furniture

Calculating weight

Weight of object = length x width x height x unit weight
- Round weight up.
- Err on the heavy side when calculating.

[A]

Airbag Operation

Lifting heavy objects
1. Calculate the weight of the object to be lifted.
2. Choose best shoring and cribbing material.
3. Choose best lifting technique; crane, airbags, pry bars.
4. Form a lifting sequence plan.
5. Lift the object in small increments equal to the dimension of the shoring or cribbing material.
6. Add cribbing progressively.
7. Never reach under the object being lifted; use tools to move cribbing.

Airbag operation
1. Connect air supply and airbags to control unit.
2. Adjust regulator to 120 pci.
3. Place airbag on solid base under object.
4. Protect top of airbag from any sharp points on object.
5. Add air to airbag and lift object slowly and carefully.
6. Watch for instability and re-adjust as needed.
7. Place cribbing under object.
8. Deflate airbag, raise airbag base and lift again.

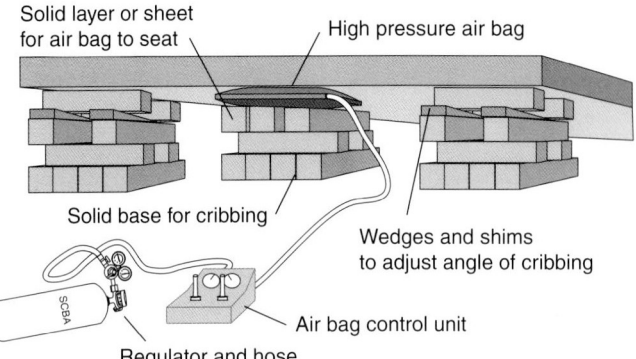

Solid layer or sheet for air bag to seat

High pressure air bag

Solid base for cribbing

Wedges and shims to adjust angle of cribbing

Air bag control unit

Regulator and hose

SCBA

T Spot Shore

Temporary shore used ONLY until a complete shoring system can be erected.

1. Survey area and determine load displacement and structurally unstable elements.
2. Clean area to be shored
3. Measure overall height of space to be shored.
   - Deduct depth of header, sole and wedges
   - Cut post to length
4. Prefabricate T spot shore in safe area.
   - Nail post to header in center of header
   - Nail plywood gusset over joint both sides
5. Place T in position with post centered under load.
   - Slide sole under post and wedge into position
   - Check shore for straightness and tighten wedges
   - Install bottom 2x4 cleat
   - Anchor header and sole to floor and ceiling

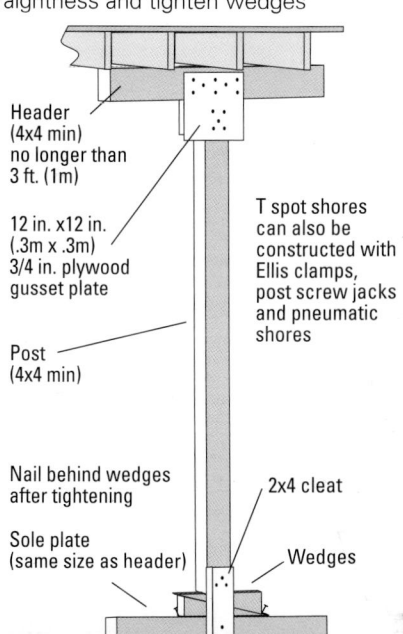

Header
(4x4 min)
no longer than
3 ft. (1m)

12 in. x12 in.
(.3m x .3m)
3/4 in. plywood
gusset plate

Post
(4x4 min)

T spot shores can also be constructed with Ellis clamps, post screw jacks and pneumatic shores

Nail behind wedges after tightening

Sole plate
(same size as header)

2x4 cleat

Wedges

Ellis Clamps

O

6000 lb. (26kn) capacity each, up to 10 ft. (3m)

1. Measure distance to be shored.
2. Each shore member should be about 3/4 of total distance. Bottom member should not be more than 7 ft. (2.1m).
3. Place two Ellis clamps on lower shore member and nail into place with 16d nails.
4. Slide upper shore member into Ellis clamps.
5. Apply a header and sole plate with gussets as shown. This configuration is a T spot shore and is very versatile and adjustable.
6. Position shore and hand tighten clamps.
7. Fit Ellis jack onto lower shore member and cam plate between cam and bottom of upper shore.
8. Jack shore to desired tightness.
9. Set clamp plates on upper shore by striking hammer lug.
10. Place a safety nail in each upper shore clamp plate.

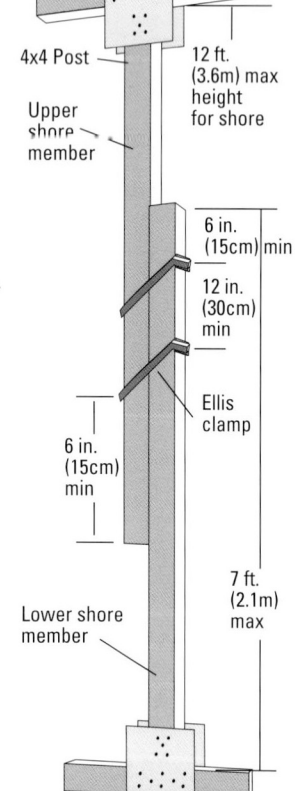

4x4 Post

12 ft. (3.6m) max height for shore

Upper shore member

6 in. (15cm) min

12 in. (30cm) min

Ellis clamp

6 in. (15cm) min

7 ft. (2.1m) max

Lower shore member

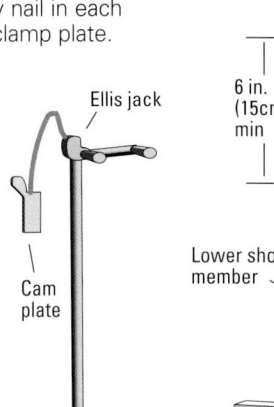

Ellis jack

Cam plate

Two Post Vertical Shore

Position header and sole across floor and ceiling joists and position posts directly under joists not greater than 5 ft. (1.5m) on center header may slope 6 in. (15cm) in 10 ft. (3m).

1. Measure distance to be shored in 3 places and use the smallest dimension.
2. Header and sole plate should be same size as post.
3. Subtract dimension of header and sole.
4. Subtract 3 in. (8cm) for 4x4 wedges.
5. Subtract 1.5 in. (4cm) for 2x4 wedges.
6. Give cut list to runner.
7. Construct shore in safe zone if possible.
8. Place posts against header flat on ground and nail gusset plates on both sides as shown (omit one gusset plate to leave space for cross brace).
9. Nail gusset plates to sole plate as shown.
10. Move shore into position and tighten wedges.
11. Add diagonal and mid bracing as shown.
12. Toe nail wedges with one nail.

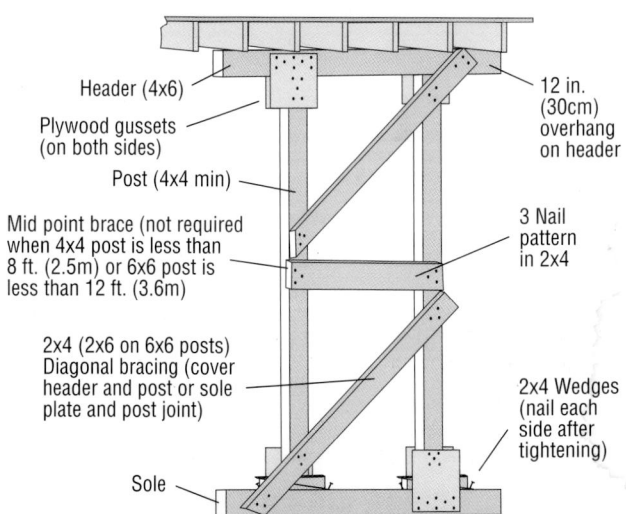

Header (4x6)

Plywood gussets (on both sides)

Post (4x4 min)

Mid point brace (not required when 4x4 post is less than 8 ft. (2.5m) or 6x6 post is less than 12 ft. (3.6m)

2x4 (2x6 on 6x6 posts) Diagonal bracing (cover header and post or sole plate and post joint)

Sole

12 in. (30cm) overhang on header

3 Nail pattern in 2x4

2x4 Wedges (nail each side after tightening)

Laced Post Shore

High capacity four post shore system that can be used to support a damaged concrete floor or a heavily loaded wood floor. Construct a pair of two post shores no more than 5 ft. (1.5 m) apart and lace together.

1. Construct two double post shore units 4 ft. (1.2m) on center from each other.
2. Tighten all wedges.
3. Measure distances for 2x6 lacing and 2x4 diagonal bracing.
4. Give cut list to runner.
5. Connect shore units into safe box as shown.
6. Use air nailer or impulse nailer for all bracing.
7. Re-tighten wedges and toe nail if possible.

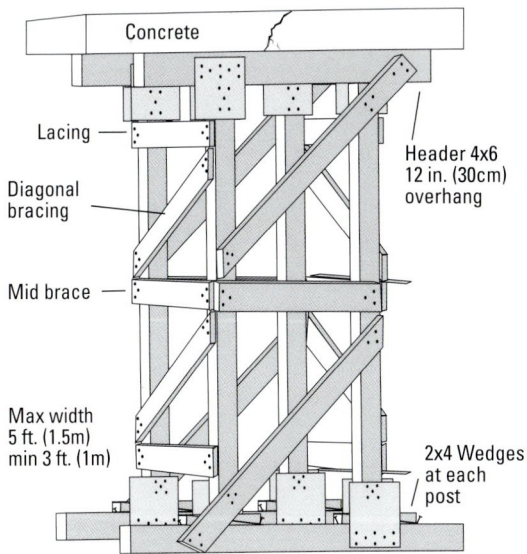

Concrete

Lacing

Diagonal bracing

Mid brace

Header 4x6 12 in. (30cm) overhang

Max width 5 ft. (1.5m) min 3 ft. (1m)

2x4 Wedges at each post

Max height = 4x width but if over 12 ft. (3.6m) high add additional mid brace

For maximum capacity use 2x6 lacing with 5 nail pattern

✓ Can be used as a refuge.
✓ Shoring concrete and steel is a technician level skill.

Alternate Door/Window Shore
Pre-constructed Door/Window Shore

1. Measure the opening in 3 places and use the smallest measurement.
2. Finished shore should be 1.5 in. (4cm) smaller than opening in each direction.
3. Construct shore in safe area as shown.
4. Install shore in opening.
5. Wedge side and shim top.

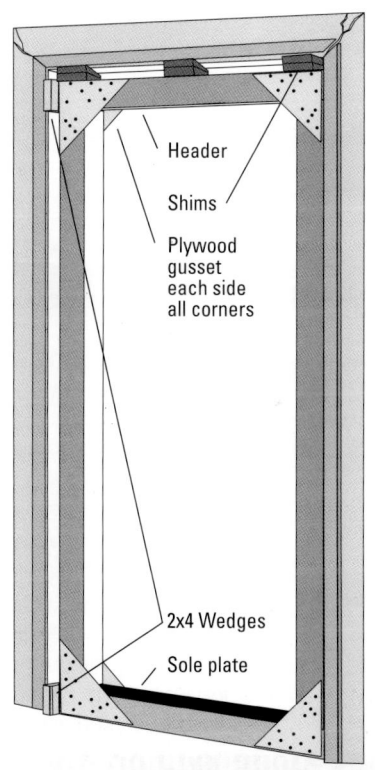

Header

Shims

Plywood gusset each side all corners

2x4 Wedges

Sole plate

Standard Door/Window Shore

1. Measure the opening.
2. Subtract 8.5 in. (21cm) for vertical posts.
3. Subtract 1.5 in. (4cm) for header and sole.
4. Use 2x4 wedges.
5. Place sole plate and tighten wedges.
6. Place header and tighten wedges.
7. Install each post and tighten wedges.
8. The header must rest on at least 1/3 of the vertical post.
9. Place 2x4 wedge-retaining cleats.
10. Install gusset plates and/or corner cleats.
11. Place diagonal cross bracing if the door is not needed for access.
12. Re-tighten wedges and secure with single toe nail to each

- Window shore steps are the same as for a doorway
- If possible, let diagonal bracing extend outside window frame to help secure shore

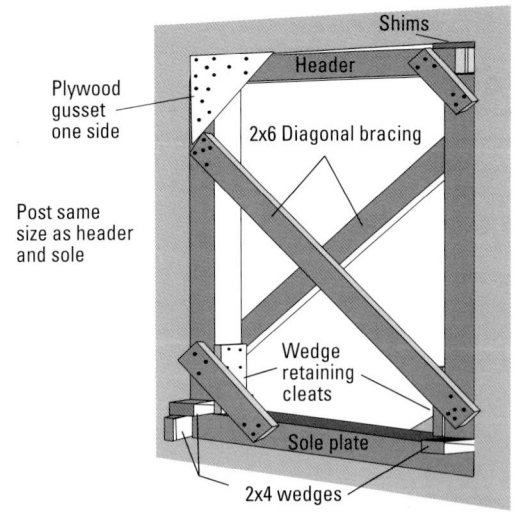

Shims

Header

Plywood gusset one side

2x6 Diagonal bracing

Post same size as header and sole

Wedge retaining cleats

Sole plate

2x4 wedges

60° and 45° Solid Sole Rakers

To calculate the length of 60° and 45° rakers

1. Measure from the top of sole plate up to the point where you want the top side of the raker to meet the wall within 1 ft. of floor level.
2. Go to the nearest foot.
3. For 45° Multiply x 17.
4. For 60° Multiply x 14.
5. Convert to inches for the required length.
   - Add 2 in. (5cm) for 45° raker cut length.
   - Add 3 in. (8cm) for 60° raker cut length.

| **Example** |
| 8" x17= 136" |
| 136"/12= 11.33' |
| 11.33" = 11'4" |
| 11'4"+2"= 11'6" |

To cut the angled ends on rakers

6. For a 45° raker, measure down 3.5 in. on one end.
7. Draw a line from the 3.5 in. (9cm) mark to the corner of the 4x4.
8. Hold a 2x4 on edge next to the line.
9. Slide the 2x4 toward the corner until one corner of 2x4 reaches an edge of the 4x4.
10. Draw a line and mark x's on the section to be removed.
11. Repeat the process on the other end, being careful to reverse the direction of cut.
12. Refer to the illustration for 60° rakers and 6x6 rakers.

Note: 30° raker is a 60° raker upside down.

60° and 45° Solid Sole Rakers

To assemble the solid sole raker shore

13. Nail a cleat onto the wall plate at the level of the raker placement (minimum 2 ft. (60cm) & 17 16d nails).
14. Place the wall plate and raker together on their side and nail the upper gusset plate onto both.
15. Place the sole plate against the wall plate (90°) and nail gusset plate on.
16. Nail the 2x6 diagonal brace from the wall/sole junction to the raker.
17. Nail the front sole gusset plate to the sole plate leaving at least 2 in. (5cm) of the raker foot exposed.
18. Flip the shore over and perform the same steps to the opposite side.
19. Stand the shore up and nail the bottom cleat into position leaving 3-4 in. (8-10cm) for wedges. (continued)

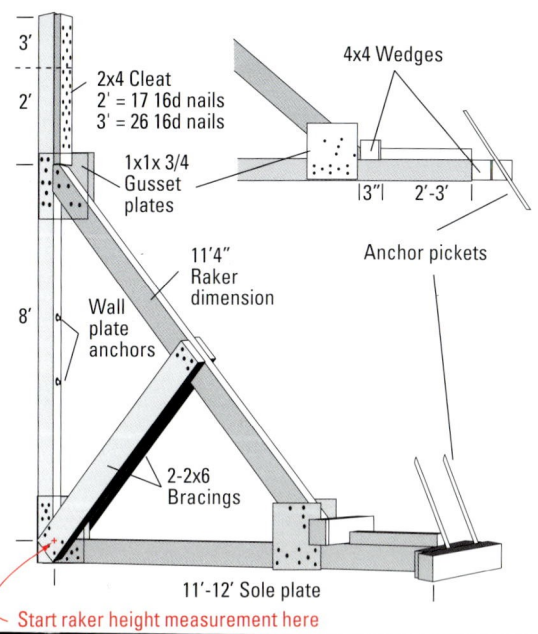

3'

2'

2x4 Cleat
2' = 17 16d nails
3' = 26 16d nails

1x1x 3/4
Gusset
plates

4x4 Wedges

|3"| 2'-3' |

Anchor pickets

11'4"
Raker
dimension

8'

Wall
plate
anchors

2-2x6
Bracings

11'-12' Sole plate

Start raker height measurement here

60° and 45° Solid Sole Rakers

20. Place one nail into the bottom raker/gusset plate contact area.
21. Move the shore into position against the wall and tighten just enough to hold into position.
22. Assemble additional raker shores and move into position (max 8 ft. (2.5m) on center).
23. Place the sole plate anchor in front of the raker set.
24. Drill 2 diagonal anchor holes in front of each raker through the sole plate anchor and into the ground.
25. Place appropriate wedges between sole plate anchor and sole plate.
26. Install the anchor pickets and tighten the anchor plate wedges.
27. Check each raker for plumb and install shims, as needed.
28. Tighten the raker wedges and toenail with one nail.
29. Connect the raker shores with 2x4 diagonal cross bracing.

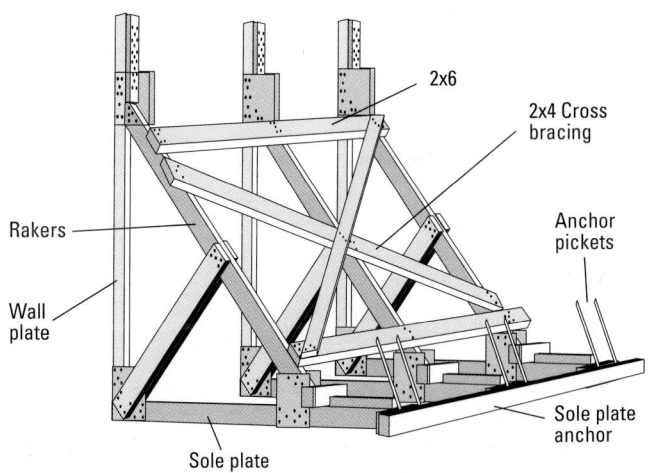

Flying Raker

The flying raker is considered a temporary spot shore to be used only until permanent shoring can be placed. Use the following steps to pre-construct the flying raker.

1. Determine height to raker insertion point on building (within 1 ft. (.3m) above or below the upper floor level).
2. Layout in safe area.
3. Cut top of raker as shown on page 131.
4. Nail cleat to wall plate and apply raker and gusset plates.
5. Apply bottom braces as shown.
6. Assemble U-channel sole plate as shown.
7. Move raker into position and dig out base for sole plate.
8. Wedge and shim sole plate as needed and anchor wall plate as shown (anchor plywood backing if used).

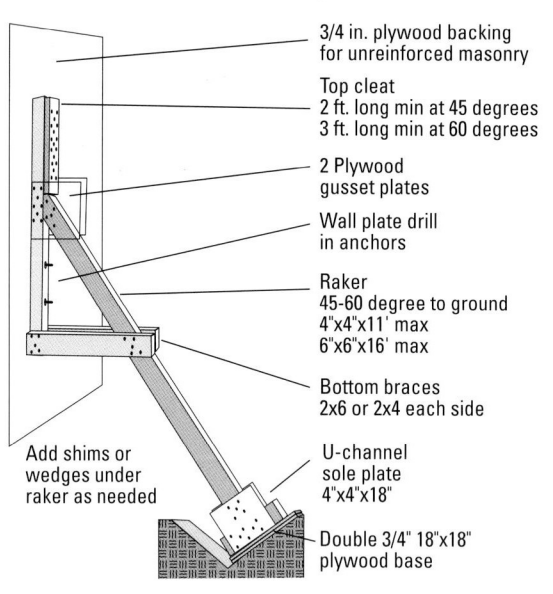

3/4 in. plywood backing
for unreinforced masonry

Top cleat
2 ft. long min at 45 degrees
3 ft. long min at 60 degrees

2 Plywood
gusset plates

Wall plate drill
in anchors

Raker
45-60 degree to ground
4"x4"x11' max
6"x6"x16' max

Bottom braces
2x6 or 2x4 each side

Add shims or
wedges under
raker as needed

U-channel
sole plate
4"x4"x18"

Double 3/4" 18"x18"
plywood base

Sloped Floor Shoring

Take off dimensions

1. The rubble or ground must resist the slope force.
2. Provide temporary shoring of sloped area.
3. Measure the total distance to be shored, divide into 1/4's and mark the 1/2 and 3/4 points for the posts.
4. The position of the long post should allow for a minimum 12 in. overhang of the header.
5. Measure from the point of contact at 90° to the slope and note the distance as B and C for long and short respectively.
6. Measure the distance on the slope between the long and short and note as A.

Layout in safe area

If more than one shore will be made, layout all shore components at the same time (3 sole plates etc.).

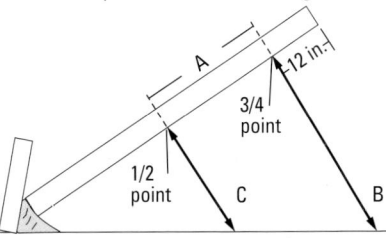

Step 1

1. Cut the header and sole to length.
2. Lay header down in layout area and mark corners.
3. Mark dimension A on the header (leave equal overhang on each side).
4. Using a square and tape measure, measure and mark distance B and C on the ground.
5. Position sole plate as shown and re-measure.
6. After double-checking measurements and position, mark the B and C point on the sole plate and mark the corners of the sole plate on the layout surface.

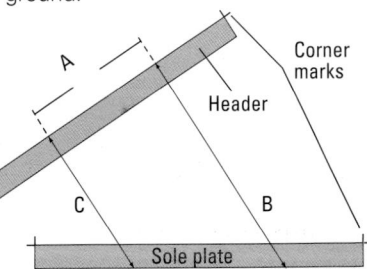

Sloped Floor Shoring

Step 2

1. Lay posts on sole plate and against header as shown.
2. Both posts must be at 90° to the header.
3. Confirm that header and sole are still in position with corner marks.
4. Mark each shore post as shown and cut.
5. Make 1.5 in. (4cm) cut at base of each post as shown below for cleat.

Step 3 Assembly

1. Place B and C posts into position and re-check corner marks on header and sole.
2. Cut 2x cleat 18 in. (46cm) long and nail B post to header.
3. Nail plywood gusset plate to bottom of C post (leave an inch of room in the inside for wedge adjustment.
4. Cut 2x6 to fit between top of C post and bottom of B post. Nail into place as shown.
5. Turn shore over.
6. Nail gusset plate to bottom of B post and sole plate.
7. Nail 2x cleat to top of C post and header.
8. Nail 2x6 brace between bottom of C post and top of B post.
9. Nail 2x cleat in front of B and C post with min. of 17 x 16d nails (leave 1.5 in. (4cm) room for 2x4 wedges).
10. Stand up shore and install 2x4 wedges and lightly tighten.

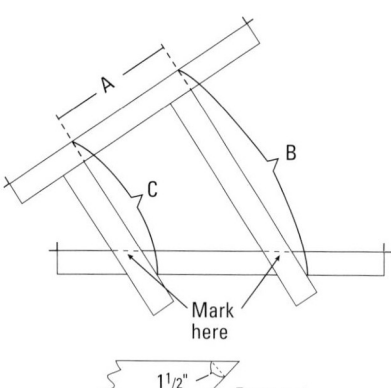

Mark here

1½"   Base cut

Sloped Floor Shoring

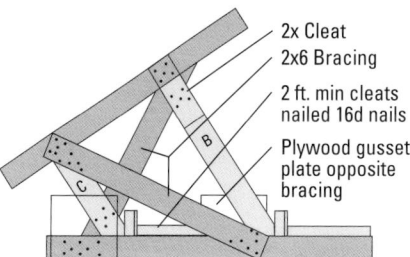

2x Cleat
2x6 Bracing
2 ft. min cleats
nailed 16d nails
Plywood gusset
plate opposite
bracing

Step 5 Install Shore

1. Move shore(s) into hazard area and place into position.
2. Place 2x4 or 4x4 wedges in front of anchor plate.
3. Place 4x4 or 6x6 anchor plate up against each set of wedges.
4. Drill 2 1/2 in. holes through the anchor plate and into the concrete or pavement in front of each header.
5. Drive rebar or the most appropriate steel through anchor plate to secure slope shores.
6. Tighten shores with anchor plate wedges first and then with post wedges.
7. Add 2x6 bracing to link all shores together.

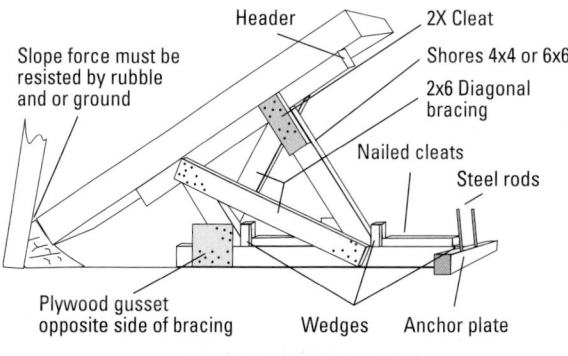

Header
2X Cleat
Slope force must be resisted by rubble and or ground
Shores 4x4 or 6x6
2x6 Diagonal bracing
Nailed cleats
Steel rods
Plywood gusset opposite side of bracing
Wedges
Anchor plate

Slope Shore Set

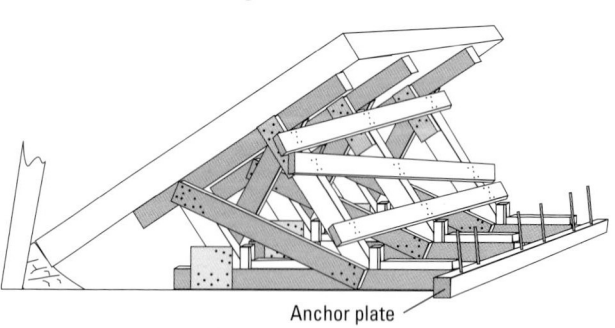

Anchor plate

Optional Vertical Post Slope Shore

If the sloping material is not connected to the structure or resisted by rubble or structural element, vertical posts should be used.

1. Take off dimension for B and C posts at 90 degrees to floor.
2. Layout in safe area if possible and construct as shown.
3. Use 2x4 or 4x4 wedged under posts.
4. Anchor header to slope with drill in rods.
5. Anchor sole plate with anchor plate.

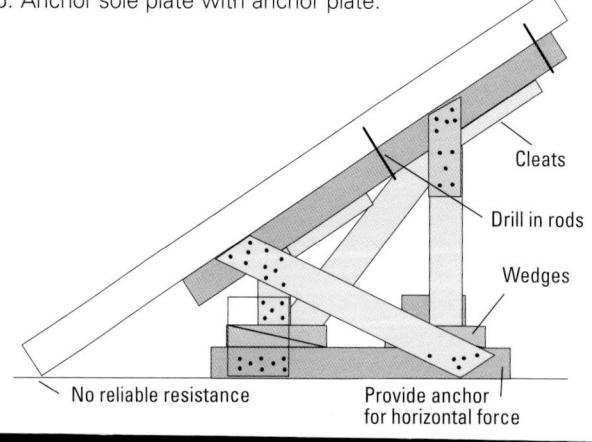

Cleats

Drill in rods

Wedges

No reliable resistance　　　Provide anchor for horizontal force

Helicopter Ops Command Checklist

Type of mission
☐ Aerial reconnaissance
☐ Med-evac
☐ Firefighting
☐ Transport personnel and equipment
☐ Does the mission include any special use activity?
  ☐ External load (longline)
  ☐ Hover sites
  ☐ Helicopter rappelling
  ☐ Single skid landings
  ☐ Two skid, power on landings
  ☐ Water rescue
  ☐ Any takeoff or landing requiring special pilot technique due to terrain, obstacle or surface condition

Base landing zone
☐ Separate radio channel
☑ LZ description and location (60x60 ft. (18x18m) day 100x100 ft. (30x30m) night)
☐ Clear approach and departure path
☐ Hazards (do not use flares)
☐ LZ security
☐ Wind speed and direction
☐ Clean LZ of any debris
☐ Headset or flight helmet for LZ control officer

Important aviation questions
☐ Is the flight necessary?
☐ Have all hazards been identified and made known?
☐ Is there a better or safer way?
☐ Is there an overwhelming sense of urgency?
☐ Be certain the following items are acceptable:
  ☐ Communications
  ☐ Weather
  ☐ Turbulence
  ☐ Personnel qualifications and abilities

A
*

Helicopter Flight Risk Score

Rescue Flight Risk Score		Total
Time		
Day	1	
Night	5	
Wind speed in knots (per pilot)		
10-15 knot steady wind = best performance	0	
0-5 knots	3	
Gusting more than 10 knots over base speed	add 5	
Gusting more than 15 knots over base speed	add 10	
Type of use		
Normal	1	
Special	10	
Load calculation (useful load carrying capacity of helo at rescue site)		
greater than 800 lbs. (360kg)	1	
600-800 lbs. (270-360kg)	2	
400-600 lbs. (180-360 kg)	3	
200-400 lbs. (90-180kg)	5	
Air temperature		
less than 80° F (27°C)	1	
80° – 100° F (27-38°C)	2	
greater than 100°F (38°C)	5	
Total score		
If score is 5-10	Lowest risk	
If score is 11-20	High risk	
If score is 21-30	Extreme risk	
If score is greater than 30	**No go!**	

Landing Zone Safety

- Always get permission to approach the aircraft from the pilot or on-board crew.
- Always approach from the front and have eye contact with pilot or crew.
- Do not go near the tail boom when the aircraft is running.
- Do not stand or take any position under the rotor tip path.

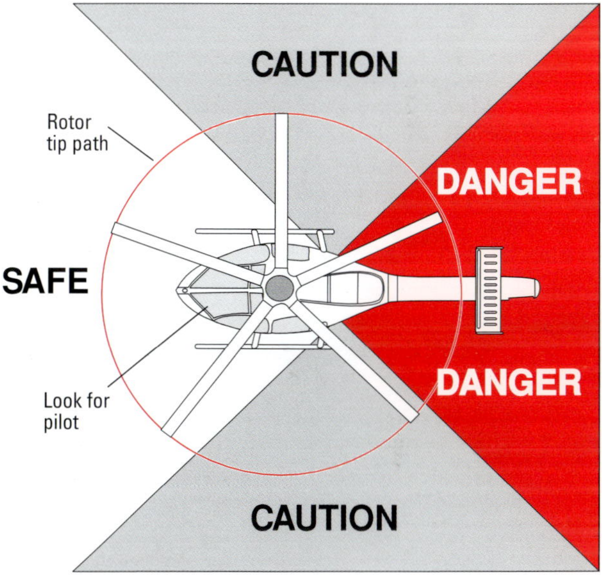

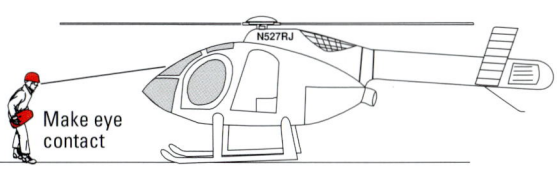

CAUTION

Rotor tip path

DANGER

SAFE

DANGER

Look for pilot

CAUTION

N527RJ

Make eye contact

Rescuer Safety

Never walk up slope!

- Do not try to talk a pilot into doing something he doesn't think is within his or the helicopters capabilities
- Average flight weight is your normal weight plus 35 lbs. (16kg)
- Eye protection should always be worn around helicopters
- The LZ control officer and the longline litter attendant should wear a flight helmet if available
- If you see a problem or if you think something dangerous is about to happen, say something!

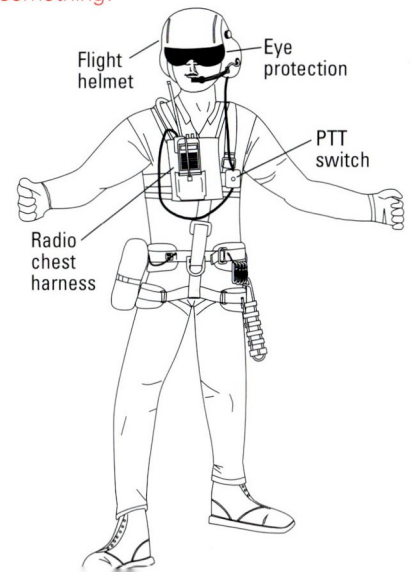

Flight helmet

Eye protection

PTT switch

Radio chest harness

Power On Insertions

One and two skid power on insertions

1. Board the helicopter and fasten seatbelts.
2. Hold gear bags securely in lap or between legs.
3. One member of the insertion team should put a headset on and establish contact with the flight crew.
4. Make sure there is nothing loose in the back.
5. Advise the crew you are ready in back.
6. Keep headset on until ship is stable at remote LZ or until crew advises you to remove it.
7. Do not remove seatbelts until clearly told or signaled to do so (over intercom or by hand signal).
8. When you receive the signal to disembark, unbuckle your seatbelt.
9. Re-buckle the seatbelt behind you.
10. Leave the gear with the other rescuer and carefully get out.
11. Remember, a gust of wind can cause the aircraft to lift off at any time. If that happens and you are mostly inside, stay in. If you are mostly outside, get off. Anticipate this happening and be ready. (continued)

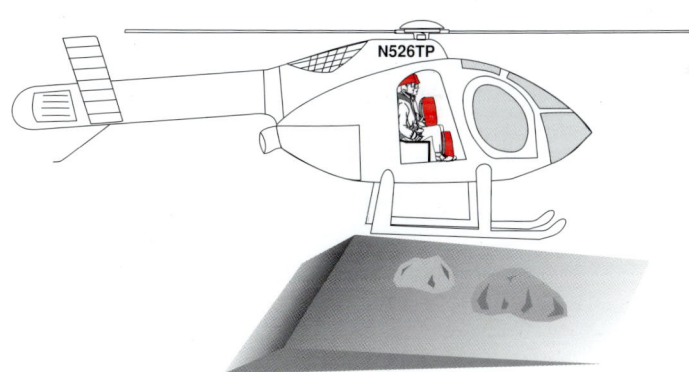

Power On Insertions

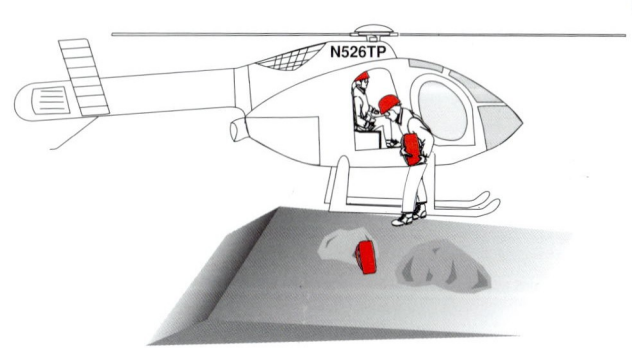

12. Do not straddle the skid.
13. Take the gear from the other rescuer and place it in a secure spot.
14. Second rescuer unfastens seatbelt.
15. Re-fasten seatbelt behind and carefully move across seats and disembark.
16. Both rescuers crouch down just in front of the skid in view of the pilot or crew.
17. Signal that you are clear with a thumbs up.

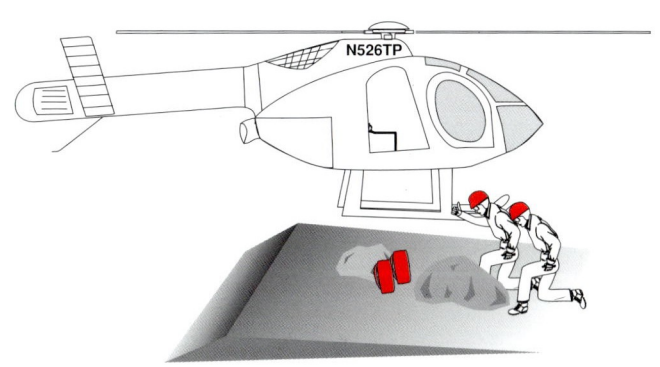

Longline Use Decision Tree

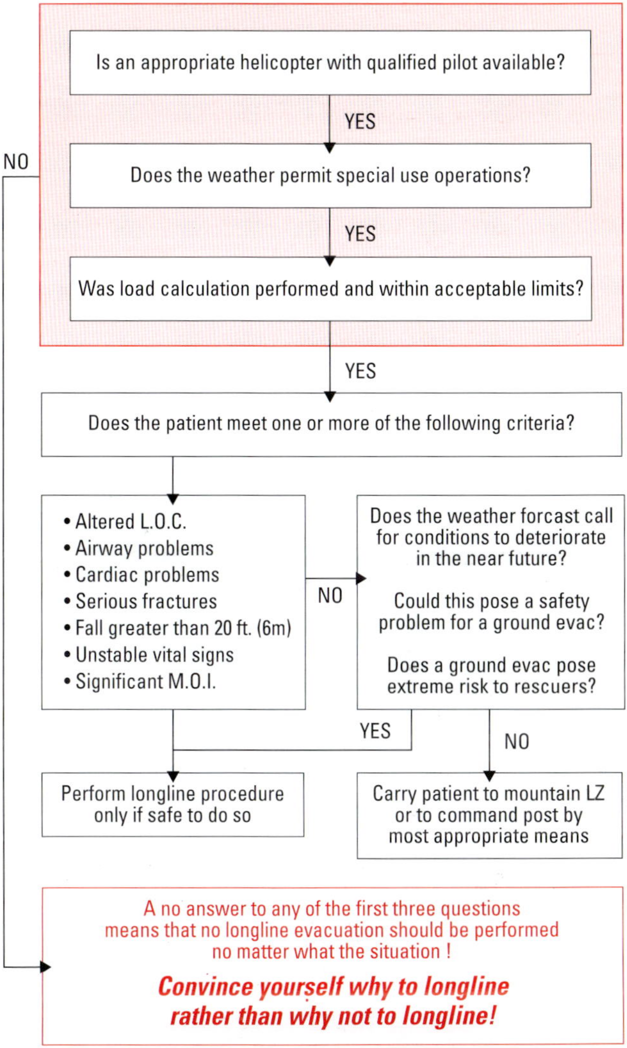

Is an appropriate helicopter with qualified pilot available?

YES

Does the weather permit special use operations?

YES

Was load calculation performed and within acceptable limits?

YES

Does the patient meet one or more of the following criteria?

NO

- Altered L.O.C.
- Airway problems
- Cardiac problems
- Serious fractures
- Fall greater than 20 ft. (6m)
- Unstable vital signs
- Significant M.O.I.

NO

Does the weather forcast call for conditions to deteriorate in the near future?

Could this pose a safety problem for a ground evac?

Does a ground evac pose extreme risk to rescuers?

YES

NO

Perform longline procedure only if safe to do so

Carry patient to mountain LZ or to command post by most appropriate means

A no answer to any of the first three questions means that no longline evacuation should be performed no matter what the situation !

Convince yourself why to longline rather than why not to longline!

Longline/Short Haul Procedures

Rigging the aircraft for longline (hot or cold)

1. Open the long line equipment bag (at aircraft cold, away from aircraft hot) and remove the belly strap.
2. Open the protective vinyl cover on the Capewell release.
3. Two technicians inspect the quick release mechanism, reset it if there is any doubt whether it is set correctly.
4. Take belly strap and longline kit to front of aircraft.
5. Two riggers lift the belly strap and remove kinks.
6. Open standard snap link.
7. Pass ends of belly strap under belly and into rear cabin.
8. Be careful of antennas.
9. Pay close attention to remove all twists from belly strap.
10. Connect belly strap at standard snap link.
11. Adjust belly strap so that quick release is near center of aircraft and so that load will hang in line with cargo hook.
12. Adjust abrasion guards to protect strap from doorsills.
13. Pull both sides of strap to middle of belly and center load D ring.
14. Position centering ring 6 in. (15cm) above load D ring.
15. Connect the backup jumper to the load D ring on the belly strap and cargo hook as shown.
16. Confirm cargo hook is set by giving gentle tug on jumper.
17. The longline safety hook should be connected to the belly strap load D ring.

Rigging the litter

1. Walk the longline bag out in front of the aircraft.
2. Connect the steel ring on the longline to the steel ring on the litter bridle with 2 steel carabiners, gates down and opposed.
3. Adjust the sandbag so that it is 4-5 ft. (1-1.5m) above the litter attachment point.
4. Check strobe function for night operations.
5. Litter attendant connects into the steel ring of the litter bridle with 2 purcells or with 2 36 in. (1m) multi loop straps.
6. A 15 ft. (4.5m) piece of webbing can be attached to the end of the litter and deployed by the attendant for a tag line.

Longline Rigging Diagram

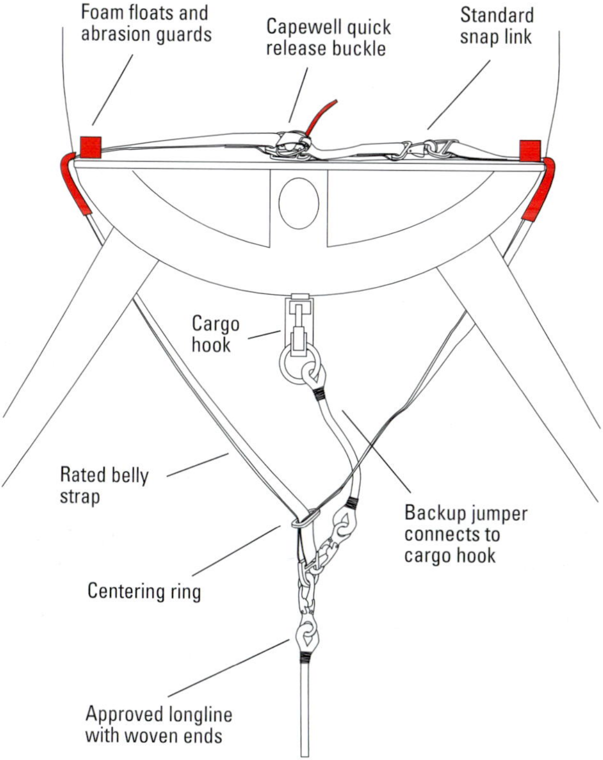

Foam floats and abrasion guards

Capewell quick release buckle

Standard snap link

Cargo hook

Rated belly strap

Backup jumper connects to cargo hook

Centering ring

Approved longline with woven ends

Some agencies may prefer to have primary weight on cargo hook and use the belly strap as backup.

Capewell® Release Mechanism

1. Inspect for any damage.
2. Seat spline on male half into notch A on female half.
3. Move release action into locked position and gently snap it into place.
4. If release action does not lock in with gentle pressure, start over.
5. Once release action is locked into place, move cable pull all the way forward in front of notch B.
6. While holding cable in position, seat end of safety cover into notch B and flip over into position.
7. To release, lift safety cover and pull cable.

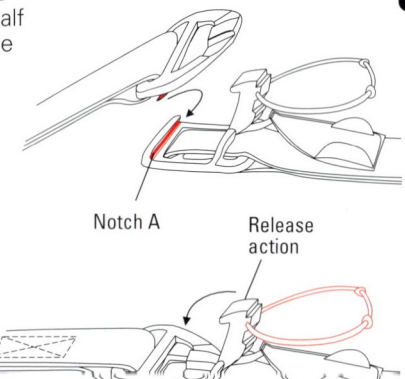

Notch A

Release action

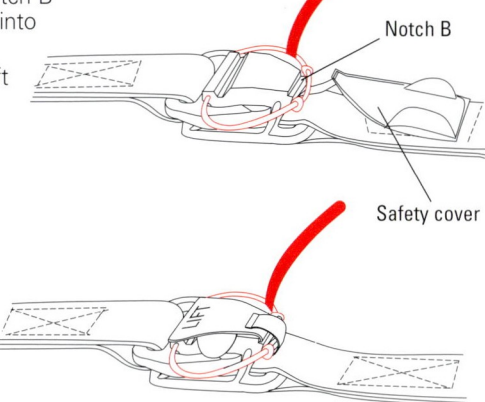

Notch B

Safety cover

Litter Rigging for Longline

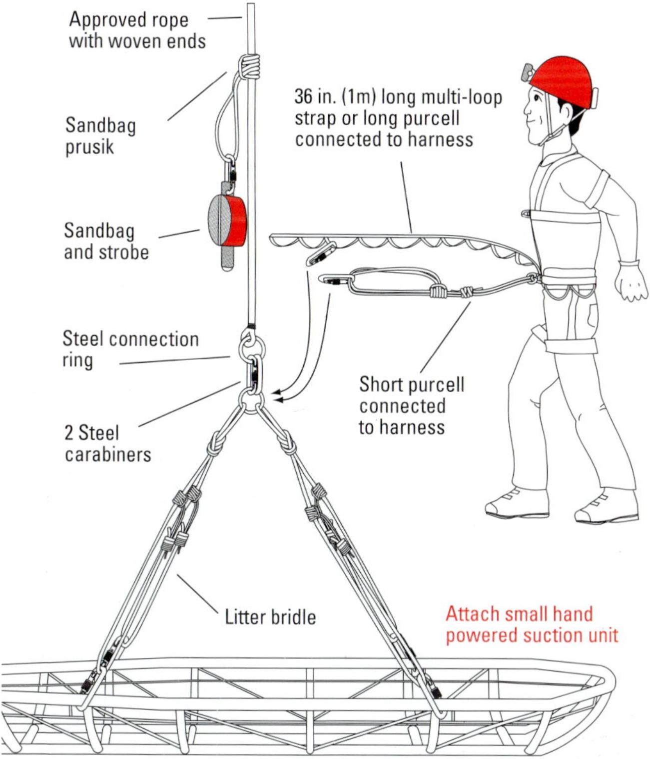

Approved rope
with woven ends

36 in. (1m) long multi-loop
strap or long purcell
connected to harness

Sandbag
prusik

Sandbag
and strobe

Steel connection
ring

Short purcell
connected
to harness

2 Steel
carabiners

Litter bridle

**Attach small hand
powered suction unit**

✔ Do not connect the longline to the rescuer and litter bridle
until completely ready for flight.

Emergency Procedures

T

Internal load
- In the event of a power failure or emergency landing, lean forward and hold lower legs tightly
- After landing, remain in the aircraft until all parts have stopped moving

External load
- In the event of a power failure while in forward flight, prepare for autorotation
- In the event of a power failure while hovering one longline length off the ground or low level hover, prepare for hard impact and take cover if possible

Sling Loading Equipment with the Cargo Hook

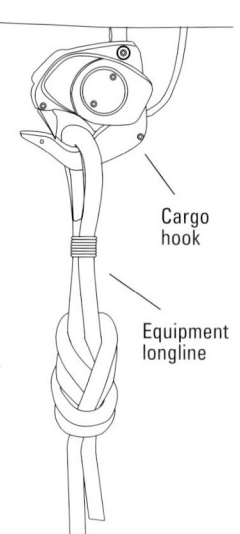

1. Use the stokes basket or cargo net to hold the gear.
2. Tie gear securely together.
3. Connect one end of the equipment longline to the equipment package with steel carabiners.
4. Connect the other end to the cargo hook on the helicopter.
5. Be sure that the bight is all the way into the hook.
6. Gently tug on the rope to confirm that the hook is set.
7. Signal the pilot that the load is secure.
8. Advise the field team that gear is on the way.
9. The helicopter will place the load near the rescue sector.
10. Disconnect load from equipment longline.

Cargo hook

Equipment longline

Rescue Medical Situations

There are certain medical situations which are common and in some cases unique to technical rescue. The following section contains brief outlines of several of these situations. As always, use common sense and follow local protocol.

Crush Syndrome

Crush syndrome should be suspected in patients who have large parts of their body (lower legs and pelvis) subjected to pressure and immobilization. In cases of severe pressure it can occur in as little as one hour but usually takes 4-6 hours to develop.

If you suspect crush syndrome, it is important to begin treatment prior to removing pressure from the patient.

Treatment (should follow local protocol when available)

- ABC's, high flow oxygen and c-spine precautions
- Cardiac monitor – watch for peaked T waves and print baseline strip
- Establish two large bore IV's and begin fluid resuscitation at 20 cc/kg NS prior to release of compression
- Consider sodium bicarb 1 mEq/kg IVP
- Consider IV dextrose and IV insulin
- Consider calcium in the event of hyperkalemia
- Contact local medical control and ask for orders for suspected crush syndrome
- Document and report suspected crush syndrome at patient transfer

Suspension Trauma

Also known as harness-induced pathology, suspension trauma occurs when an individual hangs motionless in a harness. Typically this happens when a subject's fall is arrested by their fall protection system or when a subject is overcome by exhaustion and or hypothermia.

Lack of muscle activity and the harness itself compromise venous return from the lower extremities and progressive hypotension develops leading to syncope. Syncope begins a vicious cycle of no movement and increased blood pooling, resulting in death.

Signs and Symptoms
- Light headedness
- Nausea
- Difficulty breathing
- Syncope

Prevention for Rescuers

Keep your legs moving and reposition frequently if you are required to hang in a harness for any length of time (as little as 20 minutes)

Treatment
- Advise patient to move their legs and flex leg muscles while waiting for rescue
- If unconscious, address ABC's and quickly remove them from suspension
- Keep the subject sitting up to prevent the rapid return of acidotic blood from the legs to central circulation
- Transition the subject to a horizontal position slowly over a period of 20 to 40 minutes
- Avoid rapid IV bolus and treat for potential crush syndrome at the hospital

Hypothermia

Whenever a subject is immobile in a cold environment, hypothermia can occur. Hypothermia is especially common in confined space rescue and structural collapse when access and extrication takes an extended period of time.

Moderate Hypothermia 82-89° F (27-32° C) Core Temperature

Signs and Symptoms

- No shivering
- Dilated pupils
- Bradycardia
- Decreased respiration
- A-fib
- Marked decrease in mental function

Treatment (less active/more passive)

- Careful handling (watch out for V-fib)
- Remove to warm, dry environment
- Hot packs/hot water bottles
- Warm IV fluid
- Warm humidified oxygen
- Re-warm over several hours

Severe Hypothermia <82° F (<27°C) Core Temperature

Signs and Symptoms

- Progressive decrease in metabolism can result in death
- Coma (remember, not dead till warm and dead)
- Significant hypotension
- Respiratory arrest
- V-fib to asystole

Treatment

- Careful handling (watch out for V-fib)
- Remove to warm dry environment
- Slow re-warming (very passive in the field)
- Do not give ACLS drugs until the patient is re-warmed above 86° (30°C) core temperature
- Re-warm in hospital

Appendix A

NFPA 1983, 2001 edition, pertinent definitions

Auxiliary equipment System components that are load bearing accessories designed to be used with life safety rope and harnesses including but nor limited to, ascending devices, carabiners, descent control devices, rope grab devices and snap links.

General use A designation of auxiliary equipment system components intended for use where the system could be subjected to a two person load.

Light use rope (one person) Life safety rope designed to support a one person load when in use; also can be used to support a two person load when used in systems where two ropes are used as separate and equal members. Minimum breaking strength not less than 20kn (4496 lbf.).

Light use A designation of auxiliary equipment system components intended for the sole use of the rescuer for personal escape or self rescue, or for the sole use of the rescuer in gaining access to victims.

General use rope (two person) Life safety rope designed to support a two person load when in use. Minimum breaking strength not less than 40kn (8992 lbf.).

NFPA 1670 1999 edition, operational levels

Awareness The minimum capability of a responder who, in the course of his or her regular job duties, could be called upon to respond to, or could be the first on the scene of, a technical rescue incident. This level can involve search, rescue and recovery operations. Members of a team at this level generally are not considered rescuers.

Operational The capability of hazard recognition, equipment use and techniques necessary to safely and effectively support and participate in a technical rescue incident. This level can involve search, rescue and recovery operations, but usually operations are carried out under the supervision of technician level personnel.

Technician The capability of hazard recognition, equipment use and techniques necessary to safely and effectively coordinate, perform and supervise a technical rescue incident. This level can involve search, rescue and recovery operations.

Appendix B

Standard color codes for 1 in. tubular webbing

Yellow	6 ft. (2m)	Green	20 ft. (6m)
Blue	12 ft. (3.5m)	Black	30 ft. (9m)
Red	15 ft. (4.5m)		

Standard lengths for 8mm nylon system prusiks

for use in a tandem prusik belay and in pulley systems with 1/2 in. (13mm) NFPA life safety rope (total linear measurement untied)

- Short = 54 in. (137cm)
- Long = 66 in. (168cm)

Rope Rescue equipment kit inventories

Working line kit
- 1 - large rope bag with pockets, inventory and rope log
- 1 - 200 ft. (60m) 1/2 in. (13mm) rope
- 8 - steel carabiners
- 1 - prusik minding pulley
- 4 - single pulleys
- 2 - double pulleys
- 1 - 6 bar brake rack
- 1 - anchor plate
- 3 - sets system prusiks
- 3 - yellow webbing
- 3 - red webbing
- 2 - green webbing
- 2 - black webbing

Belay line kit
- 1 - large rope bag with inventory and rope log
- 1 - 200 ft. (60m) 1/2 in. (13mm) rope
- 8 - steel carabiners
- 1 - prusik minding pulley
- 2 - load releasing hitches
- 2 - set system prusiks
- 3 - yellow webbing
- 3 - red webbing
- 2 - green webbing
- 2 - black webbing

Edge management kit
- 1 - nylon utility pack
- 2 - 33 ft. (10m) 9mm rope
- 2 - edge rollers
- 8 - steel carabiners
- 1 - knot passing pulley
- 2 - canvas pads
- 1 - pocket saw
- 3 - yellow webbing
- 3 - red webbing
- 2 - green webbing
- 2 - black webbing

Patient packaging kit
- 1 - nylon utility pack
- 4 - blankets
- 1 - set cervical collars
- 1 - red webbing
- 1 - green webbing
- 2 - black webbing
- 1 - weather barrier
- 1 - roll 2 in. tape
- 1 - litter w/face shield

Medical kit (BLS)
- 1 - nylon medical pack
- 1 - blood pressure cuff
- 1 - stethoscope
- 1 - pen light
- 4 - encounter forms
- 2 - black ink pen
- 10 - pair latex gloves
- 8 - rolls cling
- 1 - roll 2 in. tape
- 10 - 4x4 dressings
- 2 - 6x9 dressing
- 2 - formable splint
- 1 - hand powered suction
- 1 - bag valve mask
- assorted OPA's

Appendix C

Recommended rope minimum strengths

diameter in. (mm)	lbs. (kN)	use	NFPA rating
3/8 (9.5)	4496 (20)	lifeline	light use
7/16 (11.1)	6000 (27)	lifeline	light use
3/8 (9.5)	2923 (13)	floating lifeline	
1/2 (12.7)	8992 (40)	lifeline	general use
5/8 (16)	12,500 (56)	lifeline	general use

Recommended accessory cord minimum strengths

diameter in. (mm)	lbs. (kN)	diameter in. (mm)	lbs. (kN)
6mm	1620 (7)	8mm	2870 (13)
7mm	2200 (10)	9mm	3670 (16)

Recommended hardware minimum strengths

item	light lbf. (kN)	general lbf. (kN)
carabiner (major axis)	6069 (27)	8992 (40)
carabiner (minor axis)	1574 (7)	2473 (11)
descent control device	3000 (13.5)	4946 (22)
auxiliary equipment	5000 (22)	8093 (36)

Recommended miscellaneous hardware minimum strengths

item	lbs. (kN)	item	lbs. (kN)
rigging ring	25000 (111)	rigging plate	8900 (39)
prusik minding pulley	8093 (36)	standard pulley	8000 (36)
tri link (10mm)	9900 (44)	tri link (12mm)	12000 (54)
screw link (7mm)	6000 (26)	screw link (12mm)	13784 (61)
multi-loop strap	5000 (22)	anchor strap	8093 (36)
1 in. tubular webbing	4000 (17)		

Actual breaking strengths may vary from manufacturer to manufacturer.

Appendix D

Minimum Personnel Requirements

Rope Rescue
High angle: 2 TRT companies
Steep angle: 3 TRT companies
Command Post LZ: 1 Technician

Confined Space Rescue
3 TRT companies
1 Support truck
1 Hazmat company
1 Utility truck

Trench Rescue
3 TRT Companies
1 Trench support truck
1 Utility truck

Swiftwater Rescue
3 TRT Companies
2 Support trucks

Minimum Team Equipment Requirements

Confined Space
2 Multi gas monitors (calibrated)
1 Personal monitor
  per each entrant
1 Rescue tripod with
  winch or pulley system
4 Intrinsically safe lights
1 Set grounded duct for fan
2 Remote air supply carts
1 Intrinsically safe intercom kit
1 Load rated extrication device

1 Lock-out, tag-out kit
1 Intrinsically safe
  ventilation fan with cord
4 Supplied air breathing
  apparatus
1 Working line kit
1 Belay line kit
4 Life safety ropes
1200 ft. Air supply hose
1 Sked stretcher

Swiftwater
1 Inflatable rescue boat
7 Paddles
1 Fill kit
1 Line gun
1 Working line kit

1 Belay line kit
2 600 ft. (180m) ropes
2 300 ft. (90m) ropes
2 Subject PFD's with helmets
6 Throw bags

Trench
10 sheets form sheeting
20 Ellis screw jacks
4 carpenter kits
4 ground ladders
16 2x10x10's
Joist hangers
Various hydraulic speed shore
Pump can
8 4x4 's
Fluorescent ground marking paint
Patient immobilization device

2 Folding shovels
2 Square shovels
2 Round shovels
4 Five gallon buckets
10 16 ft. Utility ropes
2 Garden hoes
1 Roll hazard tape
Power saw
10 concrete stakes
Ventilation equipment
Air monitor

Suggested Resources

Bechdel, Les, and Ray, Slim 1989. *River Rescue*, 2nd ed. Appalachian Mountain Club, Boston, MA.

Brown, Michael 2000. *Engineering Practical Rope Rescue Systems*.1st ed. Delmar Learning, Independence, KY.

Federal Emergency Management Agency (FEMA) 2002. *National US&R Response System Structural Collapse Technician Training Manual*.

Frank, James A. 1998. CMC *Rope Rescue Manual* 3rd ed. CMC Rescue, Santa Barbara, CA.

Gargan, James 1996. *First Due Trench Rescue*, 2nd ed. Mosby Yearbook, St Louis MO.

International Association of Fire Fighters 1995. *Training for Hazardous Material Response, Confined Space Operations*. International Association of Fire Fighters, Washington D.C.

Phoenix Fire Department 1999. *Standard Operating Procedures*, Volume II. Phoenix, AZ.

Ray, Slim 1997. *Swiftwater Rescue*. CFS Press, Ashville NC.

Roop, Michael, and Vines, Thomas, and Wright, Richard 1998. *Confined Space and Structural Rescue*, 1st ed. Mosby Yearbook, St Louis MO.

Stoffel, Skip, and LaValla, Patrick 1998. *Personal Safety in Helicopter Operations*. Emergency Response Institute, Olympia, WA.

Vines Thomas, and Hudson, Steve 1999. *High Angle Rescue Techniques*, 2nd ed. Mosby Yearbook, St Louis MO.